Anita Barrera

Lithiaka Gems and Jewels

Their History, Geography, Chemistry and Ana

Anita Barrera

Lithiaka Gems and Jewels
Their History, Geography, Chemistry and Ana

ISBN/EAN: 9783337009526

Printed in Europe, USA, Canada, Australia, Japan

Cover: Foto ©berggeist007 / pixelio.de

More available books at **www.hansebooks.com**

LITHIAKA.

GEMS AND JEWELS.

LITHIAKA

GEMS AND JEWELS

THEIR

HISTORY, GEOGRAPHY, CHEMISTRY,
AND ANA.

FROM THE EARLIEST AGES DOWN TO THE PRESENT TIME.

BY MADAME DE BARRERA,
AUTHOR OF "MEMOIRS OF RACHEL."

"Le luxe est un des signes de la civilisation."—M. THIERS.

LONDON:
RICHARD BENTLEY, NEW BURLINGTON STREET.
1860.

LONDON:

PRINTED BY GEORGE PHIPPS, 13 & 14, TOTHILL STREET, WESTMINSTER.

The MS. of "Lithiaka" having been submitted to Monsieur Babinet, the distinguished Member of the Institut, to whom the scientific public are indebted for so many valuable contributions to its treasures, and that gentleman having done me the honor to send me his opinion, together with the reflections the perusal of the work had suggested, I deemed these few pages from his pen, would prove its best introduction to the public.

A. de B.

Paris, Nov. 12th, 1860.

A FEW WORDS ON PRECIOUS STONES;

À PROPOS OF THE BOOK ENTITLED

"LITHIAKA."

PRECIOUS Stones and Gems, may at first seem but a frivolous subject. The mention suggests a jeweller's show-rooms, or the toilet of a rich and elegant woman, whose magnificent diamonds evidence her taste and her wealth, satisfying at once the requirements of her station, and gratifying the vanity of eclipsing rivals less favored by the blind goddess. Yet science and political economy must view with no little interest—the one minerals, the optical properties of which deserve to be made the subject of deep investigation; the other, the large capital, seemingly created by caprice, but which, by the unanimous consent of all nations, claims to rank among the most refined sources of enjoyment of civilized life. It is written:—"Man shall not live by bread alone;" he lives by all that the Creator has implanted in his soul. Every power, whether religious, civil, or individual, has required precious stones—the cross of the bishop, and the parure of the millionnaire's wife, have alike demanded "Lithiaka!"

The pleasure taken by persons of taste and refinement in the possession of beautiful jewels, is not inferior to that which results from the possession of rare plate, well filled libraries, paintings by the first masters, etc., etc. Every thing that elevates the soul above the petty cares of daily life, from the elegant trifles purchased by wealth and fashion, to music and poetry, is a step forward in civilization.

Much has been said against luxury—that is, against the demand for things not indispensable to existence, and to the material life of individuals.

Certainly, the private individual who lives in a style beyond his means, and above his social position, robs, not only his own family, but society. On the other hand, the possessor of a princely fortune who exercises a sordid economy, is no less guilty of a crime of *lèse-société*, since he refuses to first-class workmen the encouragement they are entitled to from those by whom they claim to be employed.

In order to prove the fallacy of all theories of exaggerated economy, we have but to take the counterpart of that which now exists—(and that which exists, exists as a consequence of the nature of the human soul; a nature none can doubt, emanates from the great creative power). Let the sovereign assume the garb of the lower classes, and, by so doing, destroy all the manufactures that furnish us with our vegetable and animal apparel; when the aristocracy is content to wear the garments of the poor, the latter will be wearing rags. If we examine the question attentively, we shall find, that competence and luxury are but the interpretation of the divine precept—work.

"I might," says Cicero, "have remained inactive" —*licuit otioso esse mihi!* But if, for the good of his family, to gratify the natural wishes of his wife and daughters, and to make an intelligent use of the labor of his countrymen, an active and enlightened man invents machines, launches vessels, and exchanges the produce of his industry for the different precious productions of the entire world, his activity is as useful to others as to himself; and while he achieves his own welfare, he deserves credit for procuring the means of subsistence for a vast number of people who might be called his coadjutors.

James VI., complaining of the spirit of independence shown by the Corporation of London, which, according to his views, curtailed his kingly prerogatives, threatened to retire to Windsor, and deprive them of his royal presence. "Your Majesty will at least leave us the Thames," replied one of the members. It has already been said, that the various productions of the globe being disseminated in a hundred localities, it was intended that human activity should seek them, and thus make each country participate in the advantages of the whole world.

Another argument in favor of luxury is that Providence, having bestowed on us the exquisite appreciation of the beautiful in all things, the *to kalon* of Plato, has invited us to participate in all the delightful sensations that refined and delicate natures are susceptible of feeling, in the possession or at the sight of the treasures of nature and art. Among these treasures, precious stones take the first place. None of the gifts of the Deity should be repudiated.

Science is as interested as worldly luxury in the examination and study of precious stones. The study of the diamond, the sapphire, the ruby, the emerald, the topaz, the amethyst, the hyacinth, the opal, and other gems of less primitive tints, constitutes an important part of mineralogical optics, which, with Brewster, I have made the subject of deep investigation, and to which I am indebted for my admission into the Institut of France — that honor which is almost a dignity. Mineralogy, according to Haüy, calls crystals the flowers of minerals. Gems, which are the flowers of mineralogical crystals, are also the flowers of luxury; they lack not one point; their sparkling brilliancy, rich and varied tints, rarity, and costliness, fully entitle them to this graceful appellation.

The perusal of "Lithiaka" has given me great pleasure, and I may also say that I have read it with no less profit than pleasure. It is science, and science appropriated to the society of the present day. It is a work of deep and conscientious labor, devoted to that which ancient and modern times have unanimously held to be most precious in the world. So long as a sense of the beautiful, the delight in the possession of riches, the instinct of vanity, of rivalry, and of triumph in outward appearances, are inherent in the human soul, so long shall gems be held in high esteem. Hence, we may confidently predict, that their price will never be questioned, unless indeed, the Creator should see fit to effect a complete change in the instincts of humanity.

The great Haüy, taking mineralogical particles as his basis, formed all the crystals presented by nature and art. His severe geometry did not disdain to give us a book especially devoted to diamonds and gems. He points out the mode of ascertaining their authenticity, their nature, their real value. The red, the orange, the yellow, the green, the blue, the violet, and the mixed and less esteemed hues, are each examined. The weight, the colour, the double or simple refraction, are each made distinctive tokens by which to recognize the nature of gems.

The author of "Lithiaka" has perseveringly sought, in the annals of ancient and modern times, the dramatic history of diamonds and precious stones, from those that enriched the ephod of the high priest of the Hebrews to the diamonds that gleam in the crowns of modern potentates. The trade in precious stones, their geographical origin, their value, and their possession in general, have been carefully investigated. In fact, the tableau of the distribution of precious stones among all nations is a chapter of the history of universal civilization. You see Cæsar, at the banquet of Cleopatra, quaffing costly wines from great cups made of priceless gems:

> " Gemmæ que capaces
> Excepere merum."

And there too you find the carelessly garbed Consul of the epoch of the first Roman Republic:

> " Sordidus etruscis abductus consul aratris."

Both are equally anxious to enrich their country with these beautiful spoils of other nations.

The vestiges that remain of the luxury of barbarous nations—among whom the merit of gold consisted in its weight, and precious stones lacked the cutting which adds so greatly to their play and sparkle,—excite our contempt.

Precious stones are luxury amid luxury; and luxury and wealth being in point of fact the result of labor in a well organized social organization, the activity of nations may almost be estimated by the wealth they have acquired by that most moral of all elements—labor, and its faithful companions, intelligence, science, the arts, power, and all that constitutes the superiority of modern civilization. Napoleon I. was wont to say that henceforward the power of science would constitute a part of the science of power; and Watt, questioned by George III. on his labors, replied "that he had been making something very agreeable to kings—power." If we look closely at it, we shall see that he who gifts humanity with mechanical power, gifts it with intellectual power: he bestows the means of subduing physical nature. When Prometheus (the first who wore a ring, set with a stone from the Caucasus,) gave fire to man, he did not foresee that in our times fire, water, and iron, would impel cars with incredible speed; and when Thales moved light straws by rubbing yellow amber or electron on cloth, he was far from suspecting that this simple experiment, which to him was perfectly unintelligible, would one day prove the starting point of the electric telegraph. The audacious genius of man, that has compelled the earth to produce, and made fire, water, and wind labor, has at last tasked the agent of lightning, bidding it execute his mandates.

"Lithiaka," may be characterized as an Encyclopædia of precious stones. It embraces all the connections existing between gems, science, and man. The author has viewed the subject under every aspect—as a historian, a political economist, a mineralogist, a physician, and a chemist. Everything has been said that was necessary, without exceeding the limits of sustained interest. The fashionable lady will find in it the art of appreciating the riches that adorn her; the learned, a study on mineralogy; and the politician, a chapter of the history of the wealth of nations. This luxury, this crown of abundance, is a fair criterion of social prosperity; it is the *dubitur insuper* of the Gospel; in one word, it is the apex of the pyramid of prosperity, and denotes a broad and powerful basis.

I will take as a proof the least interesting view of gems: their chemical nature. The diamond is crystallized carbon; the sapphire and the ruby are crystallized clay; the topaz, the emerald, the amethyst, are, in a greater or lesser degree, merely pebbles tinged with certain metallic particles, more especially with iron, which has been called the great colorist of mineral nature. It would seem as though the mighty creative and organizing power had chosen to manifest its omnipotence, by producing the most valuable substances from the most ordinary elements. The ancient chemists asserted that the philosophical stone which was to bestow wealth and health, was to be formed of the vilest elements; and heaven only knows what they distilled with this idea. Yet modern science, by identifying carbon and the diamond, clay and the ruby, the opal and the pebble, has confirmed the assertions of the alchemists. The king of minerals, the diamond, may not withal despise the humble produce of England's sovereign coal-beds. In quantity and productive use, the fossil coal bears away the palm from the brilliant and high-priced diamond, even as the people taken *en masse*, exceed in importance the sovereign who reigns over them. Some years ago, the trade of crystallized coal, of the diamond, amounted to one million pounds sterling; while the black diamond, or coal, enriched England each year with twenty millions sterling.

At the conclusion of an article on the substances that are so greatly used to adorn the fair sex, we must return to the ladies. In an article of the *Revue Européenne*, I have had occasion to examine the manner in which providence has clothed the plants, birds, and terrestrial animals, from the equator to the icy regions of the poles. It has provided them with a garb, with furs or feathers, more or less warm, in accordance with the rigor of the climate in which they are placed. But nature has not merely sought to preserve animated creatures from meteorological influences, she has carefully provided for their adornment. Solomon in all his glory was not so magnificently clothed as a lily of the valley. All honor then to ornament, and especially to those treated of in "Lithiaka." The author has had occasion to see, in the salons of the Tuilleries, two thousand ladies covered with diamonds; an excellent study for one who was writing on precious stones. St. Augustin says "that cleanliness is a half virtue;" there is no fair lady who will not draw thence a conclusion, that a rich *parure* is an entire virtue.

BABINET,

(DE L'INSTITUT.)

CONTENTS.

PART FIRST.

GENERAL HISTORY OF GEMS.

CHAPTER I.

INTRODUCTION.

PAGE

Instinctive fondness of man for Diamonds and Precious Stones.—High honor in which they have been held in all ages and all lands.—Frequent mention of them in Sacred Writ.—Figurative use made of them by Poets.—Legends of the Talmud.—The study of Gems connected with the knowledge of many Sciences.—Portable nature of this species of Wealth.—History of celebrated Gems a source of deep interest 1

CHAPTER II.

PRECIOUS STONES AMONG THE ANCIENTS.

India the Birthplace of the Choicest Productions of Nature.—Ancient Authorities corroborative of its Wealth; Ancient Indian Poem; Quintus Curtius.—Precious Stones among the Ancient Egyptians, Babylonians, Jews, Medes, Persians, and Greeks.—Jewels establishing a link between two worlds.—Gems with the Hebrews, and with the ancient nations of South America 9

Chapter III.

PRECIOUS STONES AMONG THE ROMANS.

PAGE

Luxury imported from the East.—Height to which extravagance was carried among the Egyptians and Romans.—Cleopatra's Banquet.—Collections of Precious Stones.—Jewels of Mithridates.—Genealogies of Jewels.—Pliny.—Jewels of a Roman Belle.—Julius Cæsar.—Augustus.—Nero.—Heliogabalus.—Sumptuary Edicts 21

Chapter IV.

GEMS AMONG THE GAULS, GOTHS, AND FRANKS.

During the Dark Ages.—Treasures found in Tolosa and Narbonne.—The Toledan Cross.—Trousseau of a Frankish Princess.—St. Eloi, the Patron of Jewellers.—Jewels of Charles the Great.—Heaven-guarded Jewels . . . 29

Chapter V.

THE TWELFTH, THIRTEENTH, FOURTEENTH, AND FIFTEENTH CENTURIES.

1100 to 1500.

Renaissance of the Jeweller's art.—Abbot Suger and Louis VI.—Religious Jewels in the ancient Treasure-House of St. Denis.—Louis VII.—Grace-cup of Thomas à Becket.—Jewels of a Royal Bride in the Twelfth Century.—Extravagance in Jewels during the Middle Ages.—Jewels lost at the Battle of Poitiers.—Stone-broidered robes of Philip the Bold, Isabella of Valois, and Richard II.—Jewels of Louis, Duke of Orleans.—Magnificence of the House of Burgundy.—Important part played by Jewels in the history of the Dukes of Burgundy.—Diamonds. 38

Chapter VI.

THE SIXTEENTH CENTURY.

1500 to 1600.

PAGE

Louis XI.—Charles VIII.—Francis I.—Cardinal d'Amboise's Encouragement to Arts.—Jewels of that day.—Anecdote of the Jewels of the Countess of Chateaubriand.—Henry II. of France.—Henry VIII. of England.—Katharine of Aragon.—Anne of Cleves —Elizabeth.—Passion of that Queen for Jewels.—Jewels of Mary, Queen of Scots . 54

Chapter VII.

THE SIXTEENTH AND SEVENTEENTH CENTURIES.

Sumptuary Edicts of Charles IX. and of Henry III.—Dress of Henry IV.—Dress of Mary de Medici.—A costly Kerchief.—Extravagance of the Court during Minority of Louis XIII.—Magnificence of the Court of Louis XIV.—The fatal Bouquet.—Curious Letter of James I. —Jewels worn in Hats.—An expensive Suit of Clothes . 66

Chapter VIII.

THE EIGHTEENTH CENTURY.

1775 to 1785.

The Diamond Necklace.—An unlucky Witticism.—Blood-royal in the Hospital.—A Confidential Friend of the Queen.—Royalty in Pecuniary Difficulties.—A Friend in Need.—The Invisible Purchaser.—A Prince's I.O.U. —The Queen's Billet-*due*.—Slight difference between Buying and Paying.—Disagreeable Morning Callers.—Starting for Church and going to Prison 78

Chapter IX.

THE EIGHTEENTH AND NINETEENTH CENTURIES.

The Revolution.—The Directoire.—The Consulate and Empire.—The Restoration 104

PART SECOND.

THE GEOGRAPHY OF PRECIOUS STONES.

Chapter I.

ANCIENT FICTIONS AND MODERN DISCOVERIES.

PAGE

The Mythological Origin of many Gems.—The Diamond, the Amethyst, the Pearl, the Emerald, Lapis-Lazuli, Amber.—Ignorance, or at least silence, of the Ancients with regard to Mines of Precious Stones.—Jealousy of Ancient Traders in Gems.—The Griffin-guarded Emeralds of old.—Egyptian Mines of the Ptolemies brought to light in the Nineteenth Century.—Demon-guarded Mines of the Present Day.—Emerald Mines in the Tyrolese Alps 114

Chapter II.

GEOGRAPHY OF DIAMONDS IN THE OLD WORLD.

Diamonds, where found.—Travels of Tavernier.—Diamond Mines of India; mode of working them.—Juvenile Merchants.—Singular hiding-place for Stolen Goods.—Silent Mode of Barter.—The Koh-i-noor as seen by Tavernier. —Diamond Mines of the Island of Borneo, of Siberia, of Algiers 127

Chapter III.

GEOGRAPHY OF PRECIOUS STONES IN THE OLD AND NEW WORLDS.

Ruby and Sapphire Mines of Pegu and Ava.—Monopoly of Rubies by the Indian Princes.—The Island of Ceylon and its Gems.—Difficulty of ascertaining where the Ancients procured their Gems.—Ancient Authorities:

PAGE

Ctesias, Theophrastus, Herodotus.—Trade carried on by the Phœnicians and Carthaginians in Precious and Fine Stones.—Influence of the Conquests of Mexico and Peru on the Commerce of Gems.—Immense quantities of Emeralds brought from the New World by the Companions of Cortez and Pizarro.—Phenomenal Emerald. —The five Emeralds of Cortez.—Trying Test.—Demon-guarded Emerald Mines of Peru.—Revival of the Emerald Trade of late years 138

CHAPTER IV.

GEOGRAPHY OF DIAMONDS IN THE NEW WORLD.

Discovery of the Diamond Mines of Brazil.—Diamond Districts of Serro de Frio.—Diamond Mines of the Province of Matto Grosso.—Process of Extraction.—Former Prejudice against Brazilian Diamonds.—Fallacy of Judgment of great Jewellers in the past Century.—Brobdignagian Jewel.—All that glitters not a Diamond . . . 151

PART THIRD.

THE CHEMISTRY OF DIAMONDS AND PRECIOUS STONES.

CHAPTER I.

SCIENCE AMONG THE ANCIENTS AND MODERNS.

Opinions of Ancient Authors as to the Origin of Gems.—Aristoteles.—Theophrastus.—Avicenna.—Falopius.—Cardan.—De Clave.—Boetius de Boot.—The Diamond. —Its Hardness.—Lustre.—Refraction.—Phosphoric and Electric Properties.—Crystallization.—Mode of Testing.

PAGE

—Specific Gravity.—Rough Diamonds.—Mode of Cutting.—The Brilliant, the Rose, the Table, the Lasque.—Different Colors.—Manner of Weighing.—The Carat.—Use made of Poor Diamonds in the Arts.—Commercial Value of Diamonds.—Preference given to them over Colored Hyalines.—Superior Beauty of Diamonds when seen by Artificial Light.—Capital invested in Diamonds not subject to depreciation.—Chemical Nature of the Diamond.—Its Combustibility.—Interesting Experiments.—Attempts to make Diamonds 158

CHAPTER II.

CHEMISTRY OF PRECIOUS STONES.

Crystal.—Former Belief regarding its Composition.—Classification of Gems.—How to distinguish White Hyalines from Diamonds.—Composition of Corindons.—The Ruby.—The Sapphire.—The Amethyst.—The Topaz.—The Beryl.—The Emerald.—The Chrysolite.—The Peridot.—The Cymophane.—The Garnet.—The Hyacinth.—The Opal 180

CHAPTER III.

CHEMISTRY OF GEMS.

Hyaline Quartz, or Rock-Crystal; its composition.—The Turkois.—Agates.—The Onyx, Sardonyx, Chalcedonyx, &c.—Jasper.—Feldspath.—Lapis-lazuli.—Malachite.—Amber 196

CHAPTER IV.

PEARLS.

Diversity of Opinions as to the Origin of the Pearl.—Most probable Cause of its Formation.—Oysters inoculated with the Pearl Disease.—Chemical Composition of the Pearl.—Antiquity of the Pearl-Trade.—Ancient Pearl-Fisheries.—Pearl-Fisheries of the Present Day.—Pearl-Divers 205

CHAPTER V.

PEARLS.

PAGE

Pearl-Fisheries of South America.—Their Value when first established.—Quantity of Pearls brought to Europe.—British Pearl-Fisheries. — Bohemian Pearls. — Large Pearls.—Price of Pearls at the Present Day.—Different Colored Pearls.—Taste of the Orientals.—Goa, the Great Indian Mart for Jewels.—How to preserve Pearls.—Corruptible Nature of the Pearl. — Predilection of the Orientals for Pearls. — Pearls constitute a Portion of Regalia.—Passion of the Romans for Pearls. — When introduced in Rome.—Pearl Portraits.—Cost of a Pearl Necklace.—Cleopatra outdone.—Cæsar a Connoisseur.—An Enthusiastic Eulogy of the Pearl.—When most worn in France 219

PART FOURTH.

THE QUALITIES, PROPERTIES, AND VIRTUES, OF PRECIOUS STONES.

CHAPTER I.

MARVELLOUS PROPERTIES ATTRIBUTED TO GEMS.

Qualities, Properties, and Virtues, Natural and Supernatural, Physical and Moral, attributed in former times to Diamonds and Precious Stones.—Innate Fondness of Man for the Marvellous. — The *Diablerie* of Past Days in Better Taste than that of the Present.—Spirits in Tables, and Angels in Gems.—The Magic of the Diamond. —Opinions of a *Savant* Two Centuries ago.—The Diamond a Peace-maker and a Tale-bearer; an Antidote and a Poison.—The Diamond as an Emblem among the Ancients and the Moderns.—The Gem in the Ephod.—Marvellous Property of the Diamond 232

CHAPTER II.

OF THE VIRTUES AND PROPERTIES, SPIRITUAL AND PHYSICAL, OF PRECIOUS STONES.

PAGE

Precious Stones used medicinally.—The Five Precious Fragments.—All Gems averse to Poisons.—Gemmed Cup given by Louis XI. to his Brother.—Talisman of the Count of St. Pol.—Ring sent to Queen Elizabeth.—Innate Properties of Gems strengthened by Magic.—Ring of Louis, Duke of Orleans.—Properties of the Ruby.—Death-presaging Ring of a German Philosopher.—Male and Female Carbuncles.—Properties of the Sapphire.—Consecrated to Phœbus.—Tables of the Law.—Sapphires.—Male and Female Sapphires.—Properties of the Emerald.—The Eye-glass of a Roman Dandy.—Properties of the Topaz—of the Amethyst—of the Opal—of the Turkois—of the Beryl—of the Agate—of the Jasper—of the Heliotrope.—Great Reputation of Coral.—Consecrated to Jupiter and Apollo.—Amber.—Pearls 240

CHAPTER III.

GLYPTICS.

Intaglios.—Cameos.—Antiquity of the Glyptic Art.—Graven Gems in the Ephod and Pectoral.—Graven Talismans.—Advice of a *Savant*.—Stones preferred by Ancient Artists.—Celebrated Antiques.—Shell-Cameos . . 253

PART FIFTH.

HISTORICAL JEWELS.

Chapter I.

EMINENT PERSONAGES AND THEIR JEWELRY.

PAGE

Important part played by Jewels in the Lives of the Great.—Solomon's Magic Ring.—Talismanic Ring of Gyges.—Token Ring of Rama.—Ring of Polycrates.—Anecdote of Cæsar's Ring.—Rings sacrificed in Token of Grief.—Jewels presaging Important Events.—Nero's Armlet.—Galba's Necklace.—Galba's Crown.—Rings of Tiberius.—Crowns of Henry III., Louis XVI., and James II.—Jewels celebrated for Size and Beauty.—Three Diamonds of Charles the Bold.—His Jewelled Hat.—His Three Rubies.—Ruby in the Crown of England.—Ruby of Rodolphe II.—Of Elizabeth.—Ruby and Topaz of Runjeet Singh.—Rubies mentioned by De Berquen.—Ruby presented by the Czar to the King of England.—Emeralds of Fernando Cortez.—The Sacro Catino.—Emerald in the Temple of Boudha.—La Peregrina 260

Chapter II.

THE FINEST DIAMONDS KNOWN.

The Paragon Diamonds.—Diamond of the Rajah of Mattan.—The Orloff, or Grand Russian.—The Grand Tuscan.—The Regent or Pitt.—The Star of the South.—The Koh-i-noor.—The Shah of Persia.—History of Three Diamonds of Charles the Bold, including the Sancy.—The Nassuck.—The Pigott.—The Blue Diamond.—The Crown-Jewels of Spain and Brazil 274

Chapter III.

DIFFERENT JEWELS WORN IN ANCIENT AND MODERN TIMES.

PAGE

Nose, Chin, Cheek, Lip, and Ear Jewels.—Pistols and Poniards.—Jewels of a Daughter of the House of Alba.—Daggers of Eastern Princesses.—Military Collars and Chains of the Romans, Gauls, Mediæval Knights, and of Modera Orders.—Necklace of Penelope; of Eriphyle; of Agnes Sorel; of the Queen of Scots; of the Duchess of Berry; of the present Queen of Prussia; of the Empress Eugénie.—The Girdle an Insignia of Knighthood.—Charmed Girdle of Pedro of Castile.—Diamond Girdle of Isabel II.—Crowns.—Floral Coronals.—Gold Crowns.—Crowns Military Rewards.—Different kinds of Crowns.—Snuff-boxes.—Shoe-buckles 289

Chapter IV.

EAR-RINGS.

Their Antiquity.—Juno's Ear-rings.—Penelope's Ear-rings.—Ear-rings of the Egyptian Ladies.—Eve's Ears bored.—Fatal Ear-rings of the Israelites.—Arab Saying.—Weight of Jewish Ear-rings.—Ear-rings among the Greeks and Romans.—Extravagance of Roman Belles.—Different kinds of Ear-rings.—Ear Doctresses.—Title of the Emperor of Astracan.—The Ear-ring an Insignia of Knighthood.—Ear-jewels of the Chola Girls 308

Chapter V.

BRACELETS.—ARMLETS.

The Armlet a Token of Sovereignty in the East.—Worn by men.—Antiquity of the Bracelet.—Egyptian Bracelets.—Not an Ancient Fashion with the Greeks.—Mentioned in Holy Writ.—The Bracelet among the Moderns in the East.—The Armillæ of the Romans.—Armlets of the Sabines.—The Bracelet not worn by Girls.—Different kinds of Bracelets worn in Rome.—Armlets of the Gauls and Saxons.—Used to render Contracts binding.—Celebrated Armlets 314

Chapter VI.

RINGS.

PAGE

Earliest mention of Rings.—Mistaken idea of the Ancients. —The Ring a Symbol of Omnipotence.—Rings among the Hindoos.—In Holy Writ.—Seal-ring among the Hebrews.—No mention of Rings in Homer.—Seals, but no Seal-rings, among the Ancient Americans.—How worn among different Nations.—Rings among the Romans. —The Iron Ring.—Fable of Prometheus.—The Gold Ring.—By whom worn.—Dissensions it occasioned.—Edicts with regard to it.—Absurd length to which the Fashion was carried.—Rings on every joint.—Winter and Summer Rings.—The Pugilists' Ring.—Rings of Lawyers and Orators.—Hired as aids to Eloquence.—The Dactyliomancia, Charmed, Consecrated, and Hallowed Rings 320

Chapter VII.

RINGS.

The Sigillarius or Seal-ring.—Rings of Alexander, Sylla, Cæsar, Pompey, Augustus, and Mæcenas.—Seal-ring the Prerogative of the Wife or Eldest Daughter.—The Episcopal Ring.—The Annulus Piscatoris, or Fisherman's Ring.—The Annulus Sponsalium, or Nuptial Ring.—The Doge's Ring.—Armenian Betrothals.—A Kiss, a Ring, and a Pair of Shoes.—The Prodigal Philosopher. —The Mourning Ring.—The Gimmal Ring.—Rings given at Weddings.—Rings in the Middle Ages.—*Vie et Bagues Sauves.*—*Une Bague au Doigt*—An Arab Saying. —The Thumb Ring.—Poison in Rings.—Rings of Demosthenes and Hannibal.—Roman Lovers.—Rings as Souvenirs, Signals, Passports, Safeguards.—Devices in Rings.—A Persian Custom.—Fashion for Rings under Henry III., Louis XVI., The Directoire.—Love's Telegraph.—Letter of Pope Innocent to King John . . 334

Chapter VIII.

PAWNED JEWELS.

PAGE

Jewels a ready Resource.—Many an unsuspected *Parure* acquainted with *My Uncle*.—Diamonds of Mademoiselle * * * * *, ten years at the Mont de Piété.—Jewels a safe Investment during the Middle Ages.—Ancient Romans *au fait* in the Mysteries of Pawning.—Vitellius pledges his Mother's Pearl.—The Sand-filled Coffers of the Cid.—Henry III. pledges the Virgin, and Edward III. his best friend.—The Black Prince, Henry V., Henry VI., and Richard II.—Jewels of the Great continually travelling back and forth.—Jewels of the Duke of Burgundy pawned.—Poverty of the King of France, and rapacity of his Nobles.—Jewels of the Dukes of Orleans pledged.—Of Elizabeth of York, Henry VIII., Anne Boleyn, James VI., Henrietta Maria, and Mary Beatrice of Modena.—Napoleon I. and the Regent.—Annual Report of the Mont de Piété 353

Chapter IX.

GREAT JEWEL ROBBERIES.

Robbery of the Garde-Meubles.—Of the Diamonds of the Princess of Santa Croce.—Of Madlle. Mars.—Of the Princess of Orange 367

PART FIRST.

GENERAL HISTORY OF GEMS.

CHAPTER I.

INTRODUCTION.

Instinctive fondness of man for Diamonds and Precious Stones.—High honor in which they have been held in all ages and all lands.—Frequent mention of them in Sacred Writ.—Figurative use made of them by Poets.—Legends of the Talmud.—The study of Gems connected with the knowledge of many Sciences.—Portable nature of this species of Wealth.—History of celebrated Gems a source of deep interest.

"Quand les premiers d'une Société peuvent acheter des diamants, les derniers peuvent acheter des aliments; mais quand les premiers en sont réduits aux aliments, ou même à la gêne, il y a longtemps que les derniers sont morts de faim."—M. Babinet.

"One of the common marks of opulence and taste in all countries is the selection, preservation, and ornamental use of gems and precious stones."—Professor Tennant.

Man—and in this generic word is of course included the fairer portion of humanity—man is irresistibly attracted by light and brilliancy, while gloom inspires him with an unconquerable aversion, and utter darkness, an instinctive horror. Everything that shines and scintillates, allures and pleases him; everything sombre and dark displeases and repels him. The dawn of day reanimates his spirits, re-awakens hope and joy in his heart, vivifies his whole being. The refulgence of the noonday sun, the splendid hues with which, at its setting, it emblazons the horizon, are to him sources of rapturous, inexhaustible admiration.

B

As great, if not greater, is the charm, when to the glories of light is added the boundless wealth of color. The gay tulip, the blushing rose, the golden scarabæus, the gorgeous peacock, the infinitely varied beauties of the butterfly, the brilliant plumage of the humming-bird, all enchant the eye of man.

At night, light is still better appreciated. Necessity and contrast enhance its value. Hence man never wearies of contemplating the mild radiance of the moon, and the glistening eyes that stud the vault of heaven. On earth his gaze delights to follow their bright rays, piercing at intervals the foliage of the grove, or pouring a sheet of molten silver over the crystal waters. Even the flitting, cheating gleam of the capricious Will-o'-the-Wisp can draw him astray, and the path that is made luminous by the nocturnal host of glow-worms and fire-flies—those diminutive terrestrial stars—shortens before his weary feet.

When we reflect on this instinctive taste, we cannot wonder at the fondness of man for the diamond, which concentrates and multiplies so brilliantly the light—for the precious stones which modify, but to reproduce the most splendid hues of the rainbow.

If man possessed the power attributed by poets and painters to angels and genii, of adorning his brow with one of the stars he admires in the firmament, the celestial vault would soon be left to mourn in darkness the absence of its earth-exiled luminaries.

Diamonds have supplied the place of stars.

Men and women, sovereigns and their subjects, all have, according to their means and circumstances, possessed themselves of these, the rarest, the most precious, and most splendid insignia, in order to give greater *éclat* to their rank, enhance their personal charms, and dazzle the multitude. Monarchs wear them in their crowns, their sceptres, and their swords. The breasts of the chief-dignitaries of a court are resplendent with costly gems. Nor are women behindhand in appreciating this aid to their beauty; they wreathe them in their tresses, clasp them round their throats, their arms, their waists, decorate their bosoms, ears, fingers, ankles, and even in some lands, their very toes and nostrils with them; using these sparkling trinkets to attract attention to the charms they deem most worthy of admiration.

The high honor in which precious stones have been held from the most remote antiquity, is evidenced in Sacred Writ, where we find them used to institute the highest comparisons, to denote the highest degree of perfection.

The New Jerusalem, or The Church, was revealed to St. John under the figure of a vast and superb edifice, with a wall of jasper, and foundations of all manner of precious stones; each of the twelve doors was one pearl. (Rev. xxi. 18, 19, 20, 21.)

The breast-plate of the High Priest contained twelve stones of inestimable value,—a sardine, a topaz, a carbuncle, an emerald, a sapphire, a diamond, a ligure, an agate, an amethyst, a beryl, an onyx, and a jasper; each of which was engraved with the name of one of the tribes of Israel.

In the East, wealth was estimated by the value of the jewels a man had, more than by the value of any other kind of property.

Gems formed the chief item in the paraphernalia of eastern imagery. Infinite and very beautiful are the metaphors in which the oriental poets have used them.

In the Talmud it is said, that Noah had no other light in the Ark than that furnished by precious stones.

In the legends of the Talmud it is also said, that Abraham was very jealous of his numerous wives, and kept them shut up in an iron city, which he built for that purpose; and the walls of which were so high that the sun, moon, and stars, were never seen by its prisoned inhabitants. To supply them with light, Abraham gave them a great bowl full of jewels, which illuminated the whole city.

According to the same good authority, it would appear that one object in nature could alone be esteemed of higher value than pearls. Another Rabbinical story tells us, that on approaching Egypt, Abraham locked Sarah in a chest that none might behold her dangerous beauty. "But, when he was come to the place of paying custom, the collectors said: 'Pay us the custom.' And he said: 'I will pay the custom.' They said to him: 'Thou carriest clothes,' and he said: 'I will pay for clothes.' Then they said to him: 'Thou carriest gold,' and he answered them: 'I will pay for my gold.' On this they further said to him: 'Surely thou bearest the finest silk;' he replied: 'I will pay custom for the finest silk.' Then said they: 'Surely it must be pearls that thou takest with thee,' and he only answered: 'I will pay for pearls.' Seeing that they could name nothing of value for which the patriarch was not willing to pay custom, they said: 'It cannot be but thou open the box, and let us see what is within.' So they opened the box, and the whole land of Egypt was illumined by the lustre of Sarah's beauty—far exceeding even that of pearls."

Some men have preferred exile, torture, death itself, rather than give up favorite gems, and the case of the Roman senator Nonius, who suffered proscription, rather than cede his opal (valued at two millions of sesterces, or £16,800) to Augustus, might find more than one parallel among East Indian princes of modern days.

These beautiful productions of nature have not been favorites with the luxurious and frivolous children of wealth and fashion only. They have been studied with passionate devotion, by men whose deep researches and valuable discoveries in the domains of science have made their names beacons in the paths of learning. A *savant* of our own day* remarks, that the study of precious stones, which may seem frivolous when these are looked upon as mere ornaments, appears in another light when considered with regard to important questions of trade, and as connected with mineralogy and optics, two sciences which in the present day have made such rapid strides.

The exhibition of a collection of precious stones always proves a great attraction. Those who bestow upon them the attention to which their beauty and rarity entitle them, will be gradually led to acquire some knowledge of the geography, mineralogy, physics, chemistry, and crystallography of the countries whence commerce brings these bright productions; while the fair wearers will often find it useful to be able to judge correctly of the value of their jewels, of the cutting, setting, and even of the most advantageous mode of adapting them to their beauty.

Not the least among the merits of this species of wealth is its portability in seasons of difficulty and danger. Without having recourse to the singular hiding-place chosen by the negro, who, as Tavernier says, concealed a diamond weighing two carats in the corner of his eye, there are many ways of conveying in the smallest possible compass these ready companions of flight and exile that ensure invaluable resources for the future. The Prince Palatine, after the loss of the battle of Prague, took refuge in Holland, where he was fully able to pay for his welcome, having brought away upon his person the value of a million in jewels.

Of the high esteem in which precious stones were held by the ancients, we have already spoken. Pliny says that in gems we have all the majesty of nature gathered in a small compass, and that in no other of her works has she produced anything so admirable

* Monsieur Babinet.

Many persons content themselves with admiring the gem that glitters in the crowns of potentates, on fair brows and bosoms, or in the costly collections of amateurs, without knowing, or caring to know, the origin, the nature, or even the commercial history of that which attracts their attention. Yet the study is one of fascinating interest; and could we trace what may be termed the individual career of some gems, we should have the solution of many an enigma in the annals of nations, a farther insight into the mysteries of the human heart, and ample materials for the pen of the poet and novelist.

Wonderful indeed must be the history of some of the stones we look upon with careless, unreflecting admiration! Could we go back with them through the mist of past ages, to the day when they first gratified the vanity, or excited the envy of man, perchance in the gems that bedeck the tiara of the Vicegerant of Christ we might recognize those that were set in the ephod of Aaron; some fair sovereign of the nineteenth century may be wearing the jewel that adorned a Semiramis, or a Cleopatra; or some modern Aspasia or Rhodope may possess the imperishable trinket bestowed by an Alcibiades or Pericles on the fair hegiras of ancient Greece.

The dactylothecæ which a former faith consecrated in the temples of pagan divinities, the treasures which the piety of the Middle Ages collected in Christian basilicas, have been alike scattered and dispersed, but many, very many of the gems were beyond the destructive power of man, and could bid defiance to time itself; they exist, they keep their pomp among us, and yet we know them not; what they were in the Past, they are in the Present, they will be in the Future throughout all time.

There is something exceedingly sad to a reflecting mind in the perusal of the catalogue of a great jewel sale, in which the names of former owners are appended to the several articles in order to enhance their value. Gems are never, in one sense of the word, *second-hand.* Their value, on the contrary, is frequently increased by their antiquity, whilst their primitive lustre and beauty remain undiminished. When we are acquainted with the history of a former illustrious owner, we look with mournful interest on the gem that has survived every grief, every joy, every fear, and every hope of the bosom whose agony could not dim its sheen, to sparkle as brilliantly over some other despair! Before us is one of these catalogues containing items that in one brief sentence record the cataclysm of empires:—

> "Curious drop-shaped brilliants, forty-nine grains.—Formerly in the St. Esprit, belonging to Louis XVI.
>
> "Brilliant drops; one hundred and a half grains.—Formerly belonging to Marie Antoinette.
>
> "A magnificent Rose Diamond, sixty-three grains.—Formerly belonging to Sultan Selim.
>
> "Fine brilliant drops, one hundred and a half grains.—Formerly belonging to Joseph Buonaparte."

Four names telling fearful tales of howling mobs, of wholesale massacres, of lone vigils full of agonizing terrors horribly realized, of beheaded, deposed, and strangled sovereigns; at best, of flight, exile, and death, in a foreign land! Yet the gems bear no trace of out-poured blood; they will be prized and treasured when perchance the very existence of their possessors, the very site of the events connected with them, will be matter of controversy with future *savants!*

Yet what painful images busy fancy conjures up at the sight of jewels worn when perchance no dream of evil, sadness, or death, disturbed the beings, over whom so cruel a fate was impending. They adorned ears that listened to the strains of Glück and Piccini, or to the still more fascinating voice of the flatterer; while, concealed 'neath the veil which a wise Providence throws over futurity, the scaffold and the knife of the guillotine awaited the wearer.

Could the numerous jewels that have found their way to England since the late terrible scenes in India, bear witness to the circumstances under which they have passed from the possession of their Indian owners, we question whether vanity would prove stronger than horror, and if the European fair one would dare to deck her brow with these dearly-bought gems. The *Times'* correspondent with the army at Lucknow, had the following passage relative to these bloody trophies :—

"Ere this letter reaches England, many a diamond, emerald, and delicate pearl will have told its tale in a quite pleasant way of the storm and sack of the Kaiserbagh. It is quite as well that the fair wearers—though jewellery after all has a deadening effect on the sensitiveness of the female conscience—saw not how the glittering bauble was won, or the scene in which the treasure was trove. Indeed it is only truth to tell that most of these interesting memorials of the siege of Lucknow were bought—bargains very often—by officers from soldiers hot from plunder. And some of those officers have made literally their fortunes."

CHAPTER II.

PRECIOUS STONES AMONG THE ANCIENTS.

India the Birthplace of the Choicest Productions of Nature.—Ancient Authorities corroborative of its Wealth; Ancient Indian Poem; Quintus Curtius.—Precious Stones among the Ancient Egyptians, Babylonians, Jews, Medes, Persians, and Greeks.—Jewels establishing a link between two worlds.—Gems with the Hebrews, and with the ancient nations of South America.

"Munera præterea, Iliacis erepta ruinis,
Ferre jubet; pallam signis auroque rigentum,
Et circumtextum croceo velamen acantho,
Ornatus Argivæ Helenæ: quos illa Mycenis,
Pergama quum peteret inconcessosque hymenæos
Extulerat, matris Ledæ, mirabile donum
Præterea sceptrum, Ilione quod gesserat olim,
Maxima naturum Priami, colloque monile
Baccatum, et duplicem gemmis auroque coronam."—ÆNEID. B. 1.

THE taste for precious stones is found in the most remote antiquity of which any vestiges remain, and though in what country they were first discovered, by what nation or individuals they were first worn, cannot with any certainty be determined, many circumstances concur to point out India as their birthplace. For every luxury, every rare and beautiful production of nature to which man has attached a high price, every delight that can charm his senses, every splendour that can gratify his vanity; for perfection in plants, flowers, perfumes, animals, birds, insects, gems, pearls, we turn to the East. In Asia we find the richest and most fertile countries in the world: Asia Minor, the provinces of the empire of New Persia, from the Tigris to the Indus, Northern Hindostan, with the two peninsulas on either side of the Ganges, Thibet, and China Proper.

In these favored regions, diamonds and precious stones have been found in large quantities from time immemorial. In the mythology of the Hindoos gems play an important part; in their sacred traditions, their poems, their legends, jewels are always introduced. Vischnou is in the form of a handsome blue youth blazing with light. In one of his four hands he has a shell, in another a lotus-flower, in the third a club, and in the fourth a ring *Sudarsim*, which, with the precious stone on his breast, sends forth a light that illumines all the Vaikonta, the divine abode. The description of the city of Ayodhya*, in the poem of *Ramayana*, gives an idea of the luxurious splendour and high state of civilization of that remote age, when, "It was full of merchants and artizans of all kinds; gold, gems, and other precious articles were there in abundance; every one wore magnificent garments, bracelets, and costly necklaces."

The epoch of which the poem treats is about 2,000 years before Christ, when India fell under the sway of conquerors of a race inhabiting the high mountains surrounding India on the North, and who were far superior in civilization to the conquered.

We might be tempted to accuse the author of the Indian poem of poetical amplification in the highly colored pictures of Indian wealth presented, but that a serious historian has left us a no less charming description of this terrestrial Eden. "The very birds there," says Quintus Curtius, "learn to talk very readily; the elephants are larger than those of Africa, and their strength is in accordance with their size; the beds of the rivers are of gold, and the waters flow very calmly as though unwilling

* Capital of the province of the same name in Upper India. This description is given in the *Ramayana*, a poem narrating the conquest of India by Rama, whose wife had been carried away by a demon. The *Mahâbhârata* and the *Ramayana* are the two greatest epopees of the ancient Indian epic poems that have reached us.

to disturb their rich sands; the sea casts up on its margin an abundance of pearls and precious stones, and herein consists the greatest wealth of the inhabitants, especially since they have communicated their vices to foreigners, for this, the scum thrown off by the sea, has no other value than that which luxury gives it.

"There, as elsewhere, the minds of men partake of the climate and situation of the country; the inhabitants wear robes of linen which reach to their heels, sandals on their feet, and turbans on their heads: those who are distinguished by birth or wealth, have ear-rings of precious stones, and bracelets of gold; they do not often shave, but they are especially careful of the hair, and hold a handsome head in high esteem; they allow the beard to grow without ever cutting it, but shave the rest of the face. The luxury of their kings, to which they give the name of magnificence, exceeds that of every other nation. When the king is seen in public, he is preceded by officers bearing silver censers, and perfuming the path before him: he is borne in a litter adorned with pearls, which hang from all sides of it; he is clothed in a gown of linen embroidered with gold and purple: his litter is surrounded by armed guards, several of whom bear branches of trees covered with birds, whose singing diverts and relieves his mind when overburthened. His palace is enriched with golden columns, around which twines a golden vine, supporting silver birds. The king's residence is open to all, and while his hair is being adorned, he gives audiences to ambassadors and administers justice to his people. When his sandals are loosened, his feet are anointed with fragrant perfumes."

Of the taste of the Persians for gems, the quantity they possessed, and the use they made of them, the description of the court and camp of Darius, by the same author, bears witness:—

"The cavalry, composed of twelve nations of different

manners and armour, marched in a body, and was followed by the troops, called by the Persians, the 'Immortals.' These numbered ten thousand men, and surpassed in magnificence all other barbarians. They wore collars of gold, and dresses of cloth of gold, with jackets, the sleeves of which were covered with precious stones.

"At a short distance followed the king's cousins, numbering some fifteen thousand men; this troop, dressed like women rather than men, was more remarkable for luxurious apparel than for martial equipments. Next came the Doryphores, who bore the king's mantle. They marched before the chariot on which he appeared raised as on a throne. The sides of this chariot were enriched with gold and silver effigies of several gods, and from the yoke, which was covered with gems, rose two statues, a cubit in height, representing Ninus and Belus. Between these figures a sacred eagle of gold spread his wings as though about to soar. But all this magnificence was as nothing compared to that which the king's person displayed. He wore a purple vest, under which fell the folds of his long robe covered with gold embroidery; on this robe were wrought two golden sparrow-hawks, apparently ready to fight. He had on a girdle of gold, such as women wear, from which hung a sword, the sheath of which was covered with precious stones, so skilfully set that the whole mass of gems looked like a single stone. On his head was the blue tiara striped with white, which is the royal diadem, called by the Persians, *Cydaris*. Ten thousand pikemen, with silver pikes adorned with gold, followed the royal chariot. On each side of the king marched about two hundred of his nearest relatives, and thirty thousand foot-soldiers formed the vanguard of his army."—Book III., chapter iii.

Among the presents brought to Alexander by the Satrap Orsines, were chariots enriched with gold and silver, precious stones, gold vases of a great size and exquisite workmanship, robes of Tyrian purple, and four thousand talents in coin.

The belles of ancient Egypt were as fond of gems, forty centuries ago, as those of the viceroy's harem can be now. The jewels found in the Sarcophagi testify that the fair ones of those remote days adorned their brows with diadems of pearls, and their throats with necklaces of four rows of precious stones, and with gold collars; they wore gold bracelets and armlets set with amber, and ear-rings with three drops. The fingers of the men were loaded with rings.

That the paraphernalia of a Hebrew lady's toilet consisted of as many articles as that of one of her fair and wealthy descendants of the nineteenth century, we have the testimony of Isaiah, who threatens them with the loss of their "tinkling ornaments, their cauls, their round tires, the chains, and the bracelets, and the mufflers, the bonnets, and the ornaments of the legs and the headbands, and the tablets, and the ear-rings, the rings and nose jewels, the changeable suits of apparel, and the mantels, and the whimples, and the crisping pins, the mirrors and the fine linen, and the hoods and the veils." In this enumeration jewels figure largely.

Of the fondness of the Phœnicians for precious stones, many passages in the Old Testament give evidence. The magnificence of the Tyrian lords may be judged of, when we find no less than nine precious stones are mentioned as being worn on their garments; these were the sardius, the topaz, and the diamond, the beryl, the onyx, and the jasper, the sapphire, the emerald, and the carbuncle.

In Sparta,—whence the laws of Lycurgus had banished all the arts of refinement, where elegance and luxury were looked upon as crimes, where the inhabitants fed on the traditional black broth, and ignoring the comforts of a couch, stretched their hardy limbs on reeds, within dwellings built of rough boards, and wore garments that were never changed in any season, and the cleanliness of which was at all times questionable,—in Sparta, it need scarcely be said, that gems were unappreciated. But the various Grecian States differed widely from each other in their institutions, laws, manners, opinions, and tastes; the Athenians were as polished, as elegant, luxurious, and fastidious, as the Lacedemonians were coarse, rude, careless, and uncultivated. The fair Athenians, whose sole study was the art of pleasing, whose sole ambition was to excite admiration, to win love, to fascinate, spent half their time at the toilet. The Spartan damsel, on the contrary, employed her hours in such exercises as were calculated to develop to the utmost her physical powers, and to annihilate every feminine grace. By a singular contradiction, the frail fair ones who, in Athens, exercised so great an empire over the minds of the noblest and wisest of her sons, were not allowed to appear in public adorned with valuable jewels; while in Sparta this class of women was the only one permitted to wear them; trinkets of all kinds were there prohibited to honorable matrons and maidens.

Nor is it only in the earliest records of the Old World that we find traces of man's love for gems, of his delight in precious stones. In the South American traditions of ages long gone by, in the vestiges of Mexican and Peruvian antiquities, and even amid the ruins of once flourishing cities, which the perseverance of modern travellers has brought to light from beneath the detritus of primeval forests, are found ample evidences that this taste prevailed no less in the Western Hemisphere. Indeed, jewels go far to establish the analogy found by some *savans*, between the extinct races of South America and the Jews of the age of Solomon, and prove the hypothesis which gives to both the same origin. With both, the ensigns of sovereignty and power were the same, or nearly so; with both, gems were favorite symbols and terms of comparison.

The following extract from a yet unpublished work* will show the similarity of their customs and religious rites in this respect:—

"The ephod, the breast-plate, the mitre, the ouches, the girdle, and pontifical vestments of the Hebrews, were of the same shape as the ornaments of the Mexican pontiffs, and were in like manner richly embroidered and studded with pearls and precious stones.

"Besides the numerous sculptured and painted representations of the ancient Mexican pontifical ornaments, entirely similar to those which are enjoined to the Hebrew priests, in the 28th chapter of Exodus, we again meet with these costumes broidered with precious stones and pearls, in the Bodleian MS., in the ancient Mexican paintings, in the Library of Dresden, in the Oxford Library, and in the Mendoza Collection.

"The ensigns of power were the same with both nations, and consisted of crown, bracelets, sceptre, sandals, and royal mantle.

"The royal crown of the Mexicans and Hebrews bore a greater resemblance to the sacerdotal mitre than to the crown of the western sovereigns. The American Copilli and bracelets are represented in the 57th plate of the Mendoza Collection. In several other paintings executed on maguey-paper, we have all the different articles of the royal dress and ornaments of the ancient pontiffs — the sovereigns of the Anahuac—such as diadem, sceptre, mantle, bracelets, girdle, sandals, all richly embroidered with gold, and set with precious stones.

"The seals pendant from the arm, and the breast-plate, are entirely of Hebrew origin, and were worn in the same manner by the Mexican sovereigns. This is clearly expressed in the Scriptures:—

* Zerrissene Bluetter aus dem Buche den Americanischen Urvölker-Geschichte von Tito Visino.

'And thou shalt bind them for a sign (signet, seal) upon thine hand (arm), and they shall be as frontlets between thine eyes. Deut. vi. 9.

'Therefore shall ye lay up these my words in your heart and in your soul, and bind them for a sign (signet, seal) upon your hand, (arms,) that they may be as frontlets between your eyes.' Deut. xi. 18.

"This was also a Mexican custom, as we know by the testimony of Cortez, Bernal Diaz, Sahagun, Torquemada, and others, as well as by the ancient paintings on maguey-paper, and the relics of plastic effigies. The imperial ornaments of Montezuma differed little from those of Moquitucix. The Incas of Peru also made use of the Quipos—the omnipotent Pschent of the Egyptian Pharaohs—as sacred ensigns of the royal dignity.

"We have Mexican seals and rings bearing the constellation of the pisces engraved on fine stones. The ancient Mexicans, like the Hebrews, awaited the coming of their Messiah, the 'Crusher of the Serpent,' whose advent was to occur during the conjunction of Jupiter and Saturn in the Pisces, the protecting sign of Syria and Palestine.

"According to the cosmography of the Quichés, the second king of that powerful American nation was called 'Cocavib,' *i.e.* magnificent ornament. His wealth in jewels was immense.

"Another ancient king of the Quichés was called 'Cuvatepech,' *i. e.* seven signs. 'Noh,' signified a sign of the ancient zodiac, and was an emblem of reason, intellect, wisdom, and learning. King Yucub-Noh-Cuvatepech wore this sign engraved on a Smaragdus, or emerald, as the most precious ornament, and fittest symbol of his rank.

"Votan — the heart of heaven — was represented in his principal temples, symbolically, by an enormous emerald, cut into the shape of a feathered adder. His name in heaven was 'Chalchiluclitl,' which signifies, the precious stone of endurance and self-sacrifice.

"The emerald was as sacred a stone among the Israelites as among the American Indians; the stone set in the seals which the pontiffs of both nations wore on their arm, and in the large ring on the first finger of the right hand, was an emerald.

"The superstitious zeal of the first missionaries, destroyed many precious articles made of emeralds, the stones sacred to Votan; the beautiful gems were ground into dust, which was dispersed to the four winds, as remnants of execrable heathen rites.

"The ancient South Americans, among other offerings to the deity, brought gold, silver, jewels, pearls, and other precious things. This was customary with the Hebrews also. (Exodus xxxv.)

"The Assyrian conquerors who pillaged the temple of Solomon, found there an immense booty in jewels and gold and silver vessels. The Peruvian and Mexican temples, especially that of Verachocha-Pachacamac, in Cuzco, offered similar inducements to the devastating soldiers of Pizarro.

"The deist Netzahualcoyotl, King of Acolhuau, the Solomon of the Anahuac, in his sacred songs, so full of patriarchal warmth, in honor of the Supreme Being, compares the sun, the symbol of the Eternal Divinity, to a diamond with a thousand facets. This comparison to a gem which the Mexicans did not possess, is one of many instances in their Sacred Book, the Songs of Netzahualcoyotl, in which Hebrew comparisons are used, and things are mentioned of which the Mexicans could only have an idea from tradition; thus affording another proof of their Jewish origin.

"In the beautiful exhortations of a Mexican mother to her daughter, preserved for the admiration of all ages by the pious missionary, Fray Andres de Olmos, is the following :—'And thy father has polished and rendered thee brilliant, even as a precious emerald, that thou mayest appear before the eyes of the world as a jewel of perfect virtue.'

"Huehuetlapallan and Huehuetollan—the old Tlapallan and the old Tollan—were the provinces whence the ancient Mexicans drew their fine colors and an abundance of precious stones.

"The ancient Mexicans excelled in the art of cutting fine stones, attaining a perfection that European artists could not surpass with their steel-tools. The Mexicans had not the help of steel or iron, the use of which was unknown to them; they accomplished their work by means of other hard stones, or the dust of stones.

"The Chinese assert, that the great continent of 'Fusang,' situated, according to their cosmographies, at 22,000 *Lé* east of Japan, has been known to them these 4,000 years, and that, at different epochs they had sent colonies to people that vast territory. Infinite traces of this origin are, in fact, discerned by the archæologist in the apparent confusion of races in America. If, for instance, we examine the symbolical figures of the last four months of the Mexican year, we have no difficulty in recognizing three of these, of Panquetzalitzli, Tititl, and Izcalli, a perfect Chinese type, the rude costumes studded with pearls and precious stones, of that stationary civilization, such as the 'Child of the Sun,' the Emperor of the Celestial Empire, wears at the present day.

"Fray Francisco Nunez de la Vega, archbishop of Chiapas, in one of his frequent visits to his diocess, discovered the cave, called the 'Dark House,' in which the treasure, the Teoamoxtli, and the sacred tapers of Votan, were kept before the conquest. This treasure consisted in large clay and stone vases, adorned with emeralds and other precious stones of immense value, and in other rich articles of various kinds. The spot where these royal valuables are now concealed has baffled all the efforts made to find it; these riches have, since the reverend man's visit, again become invisible, at least in the form they then bore. The clay-vessels, the writings, paintings, and mummied dantas, *i.e.* tapers, as well as the great store of copal and incense, have been burnt and otherwise destroyed; but that the gold and silver articles and precious stones are still extant, there is no doubt."

Whatever conclusion the reader may draw from the above extracts in favor of the system of the author, it is very certain that gems were held in high honor in both hemispheres, and that the uses to which they were applied, and the religious convictions connected with them in all parts of the world, present a similarity that is difficult to account for, if we refuse to admit the universal origin of all Caucasian races on both continents.

The Peruvian nobles who attempted to defend their country against the Spaniards under Pizarro, wore helmets adorned with jewels, presenting in this particular a singular resemblance to the European knights of the Middle Ages, whose basnet was similarly enriched, and often by the hands of some fair lady.

Warriors with the Roman helmet, which was also that of the mediæval knights, richly ornamented with jewels and plumes of feathers, are seen on the bas-reliefs of Yucatan, Chiapas, Guatemala, and in part of Northern Mexico, on the great medallion, the Temalacatle, in the fresco-paintings of Mitla, and in Chichen-Itza.

The Incas claimed a monopoly of jewels for themselves and the nobles on whom they condescended to bestow them; yet when the first Spanish expedition landed at Tacamez in the province of Quito, they found the inhabitants of both sexes wearing jewels of gold and precious stones.

The ancient Mexicans and Peruvians, though passionately fond of brilliant ornaments, did not possess a great variety of gems. They had an abundance of pearls brought from the Gulf of California, fine turkoises, amber, malachite, grünstein, itztli, (a kind of obsidian-stone, of which ornaments were made similar to the jet ones of the present day,) and different kinds of agates. Of the hyaline corindons, however, they had but one, the emerald, which they called chalchivitl. This, as we have seen, they held in high honor. In the chief hall of justice, called the "Tribunal of God," was a throne of pure gold studded with turkoises and other gems. On a pedestal opposite the throne was placed a human scull, crowned with an enormous emerald, of a pyramidal form, over which nodded an aigrette of gay plumes and gems. Over the throne was a canopy of bright feathers, the centre of which sparkled with gold and jewels.

The mantle of Montezuma was fastened by a rich clasp of emeralds; a profusion of emeralds of great size and beauty, set in gold, adorned his person.

The Tlascalans wore helmets adorned with plumes and gems; they ornamented even their shields with jewels.

At the entertainments given in honor of the deceased Incas, "such a display" says an ancient chronicler, "was there in the great square of Cuzco, on this occasion, of gold and silver plate and jewels, as neither in Jerusalem, Persia, or Rome, nor in any other city of the world, was ever seen."

CHAPTER III.

PRECIOUS STONES AMONG THE ROMANS.

Luxury imported from the East.—Height to which extravagance was carried among the Egyptians and Romans.—Cleopatra's Banquet.—Collections of Precious Stones.—Jewels of Mithridates.—Genealogies of Jewels.—Pliny.—Jewels of a Roman Belle.—Julius Cæsar.—Augustus.—Nero.—Heliogabalus.—Sumptuary Edicts.

"Nil non permittit mulier sibi, turbe putat nil,
Quum virides gemmas collo circumdedit, et quum
Auribus extensis magnos commisit elenchos."—JUVENAL.

THE taste for splendid apparel, for rich ornaments, and consequently for precious stones, was imported from Asia into Greece, and thence to Rome. The Greeks being the nearest neighbours to the Asiatics, were the most exposed to the blandishments of luxury,—the Siren which, after civilizing, was at last to overthrow the Roman Empire.

The gorget of Alexander covered with precious stones, his mantle embroidered with gold and gems, prove that even the Macedonian hero shared the taste of the effeminate Persians. Among the Romans luxury dates in reality from the conquest of Macedonia by Paulus Emilius. It was on the occasion of the triumphs of this consul that the magnificent spoils of Greece were for the first time exhibited in Rome.

The birth-place of the fine arts, subdued by the victorious armies of Rome, refined the taste and polished the manners of her invaders. The gold, silver, silks, perfumes, pearls, precious stones, purple and scarlet dyes of the voluptuous regions of the East, the torrid African zone, and even the frozen North, gathered by commercial Carthage within her walls, taught the stern Romans new tastes, new delights, new wants, and rendered them as insatiable of wealth as they were of dominion and glory.

The subsequent acquisitions of Pompey and Lucullus, and the return of their armies, filled Rome with so many of the luxurious fashions and customs of the East, that, by many authors, the commencement of Roman luxury dates from that epoch. The Romans soon surpassed in extravagance, if not in good taste, the nations from whom they borrowed their novelties. The passion for precious stones was carried to an extravagant height.

A collection of gems bore in Rome the foreign name of dactylotheca. Scaurus, the son-in-law of Sylla, was the first Roman who possessed one, and this was probably formed from the spoils acquired by his father-in-law. For a long period there was none other until Pompey the Great consecrated, among other gifts, in the capitol, that which had belonged to Mithridates, and which was far more valuable than that of Scaurus. Besides rubies, topazes, diamonds, emeralds, opals, onyxes, and other gems, extraordinary for size and brilliancy, the dactylotheca of this, the most wealthy and luxurious of the princes vanquished by the Romans, contained a number of rings, seals, bracelets, and gold chains of exquisite workmanship.

Splendid as was this display, it was far outdone by the marvels of nature and art that were exhibited on the occasion of Pompey's triumph, among which the following were the most important:—A chess-board, with all its pieces, of gold set with precious stones; thirty-three crowns of pearls; the famous golden vine of Aristobulus, estimated by the historian Josephus at five hundred talents (2,400,000 francs); the throne and sceptre of Mithridates; his chariot glittering with gold and precious stones, and which had been the property of Darius. The conqueror himself appeared in a mantle embroidered with gold and gems, which was said to have been Alexander's. After these valuables had dazzled the eyes of the Roman people, the arms of Mithridates, the splendour of which surpassed any thing hitherto seen, were paraded. The diadem and scabbard of the vanquished monarch, each a perfect mass of magnificent gems, did not appear in the procession; they had been stolen; the scabbard alone had cost 400 talents (£76,400).

Cæsar, following the example of Pompey, consecrated to Venus Genitrix six dactylothecæ, and Marcellus, son of Olympia, one to the Apollo Palatine. Augustus offered, in the temple of Jupiter Capitolinus, in a single day, a weight of sixteen thousand pounds in gold ingots, and precious stones to the amount of ten million sesterces.

The description in Lucan's Pharsalia, of the banqueting hall in which Cleopatra feasted Cesar, would seem the mere coinage of a poet's brain, were those gorgeous splendours not confirmed by the testimony of sober history. Columns of porphyry, ivory porticos, pavements of onyx, thresholds of tortoiseshell, in each spot of which an emerald was set, furniture inlaid with yellow jasper, couches studded with gems, met the bewildered gaze of the laurel-crowned Roman, while his heart was enthralled, his judgment subdued, by the beauty of the royal hostess, whose charms were enhanced by rich spoils brought from the Red Sea, and on whose brow glittered the treasured gems of a long line of Pharaohs. With such adjuncts, such an *entourage*, no wonder the Eastern Circe obtained an easy triumph over an Anthony and a Cæsar, great masters in the art of war, but semi-barbarians compared with the luxurious children of this land of refined voluptuousness.

Once introduced into Rome, luxury made rapid strides. Furs from Scythia and carpets from Babylon, amber from the shores of the Baltic to the Danube, and precious stones, silks, and aromatics from the East, were purchased with the silver, or exchanged for the gold of the Empire. The annual loss of this trade was computed at eight hundred thousand pounds sterling, yet disadvantageous as it seems, the produce of the mines abundantly supplied the demands of commerce*.

Notwithstanding the edicts by which he endeavoured to curb the follies of others, Cæsar was himself an indefatigable collector of precious stones, chiselled vases, statues, pictures, etc., especially when they had been the work of famous ancient masters. The quantity of gems the Cæsars had at their command must have been enormous. Caligula built ships entirely of cedar with sterns inlaid with gems. These were probably fine stones such as the onyx. The Emperor's mantle was heavy with precious stones and gold embroidery; and Incitatus, his favorite horse, was covered with purple housings, and wore a pearl collar.

In the golden house of Nero, the pannels were of mother-of-pearl enriched with gold and gems. At the Great Games instituted by this Emperor, as many as a thousand lottery-tickets were daily thrown to the people; the prizes consisted of quantities of rare birds of various kinds, corn, gold, silver, robes, pearls, precious stones, and pictures; during the last days, even ships, houses, and lands.

But it was under the reign of the Antonines that luxury was carried to its greatest height. Luxury in edifices, gardens, furniture, banquets, and dress, found historians to panegyrize or satirize them as far back as the reign of Augustus; but Pliny was the first writer who spoke of precious stones.

When the mania for jewels reached its climax, it was not enough for the vanity of these masters of the world that their trinkets were valuable on account of the work-

* Gibbon: *Decline and Fall of Rome.*

manship and beauty of the gems; they were not content unless they could boast of possessing such as had an illustrious origin. A ring, a vase, a string of pearls, or a cameo, had its genealogy, and was traced back to a Cleopatra, an Anthony, or some other illustrious owner. This vanity forms the text of one of Martial's epigrams.

Men and women vied with each other in their fondness for jewels. Pliny indignantly relates that women, not content with wearing gold on their heads, arms, tresses, and fingers, in their ears, and around the corsage of their tunics, yet wore pearls on their bosom, in the dead of the night, that, even in their sleep, they might be conscious of the possession of inestimable gems.

He complains, moreover, that they wore gold on their feet, thus establishing, between the stola of the matron and the plebeian tunic, a sort of feminine equestrian order. This was but a trifling piece of extravagance, however, when compared with the whim of the Empress Poppæa, who caused her mules to be shod with gold.

Indeed, moderation could scarcely be expected from the wives of the patricians who had subdued empires, made kings their tributaries, and reigned as sovereigns over the wide domains wrested from surrounding nations to be provinces of Rome. "I have seen," says Pliny, "Lollia Pauline, who was the wife of the Emperor Caligula, and this not on the occasion of a solemn festival, or ceremony, but merely at a supper of ordinary betrothals, I have seen Lollia Paulina covered with emeralds and pearls. arranged alternately, so as to give each other additional brilliancy on her head, neck, arms, hands, and girdle, to the amount of 40,000 sesterces (£336,000 sterling), the which value she was prepared to prove on the instant by producing the receipts. And these pearls came, not from the prodigal generosity of an imperial husband, but from treasures which had been the spoils of provinces. Marcus Lollius, her grandfather, was dishonored in all the East on account of the gifts he had extorted from kings, disgraced by Tiberius, and obliged to poison himself, that his grand-daughter might exhibit herself by the light of the *lucernas* blazing with jewels."

The censorious naturalist tells us, that a Roman belle would no more have been seen abroad without her jewels, than a consul without his fasces.

The Greek and Roman jewellers had varied the form and style of ornaments to such a degree, that, according to archæologists, our most skilful modern artists are merely copyists or imitators. The works that treat of the jewellery of the ancients, furnish inexhaustible repertories to those who explore their scientific depths. Diadems, necklaces, ear-rings, bracelets, rings, pins, brooches, clasps of all shapes and dimensions, surmounted with busts, statuettes, animals, birds, insects, flowers, etc., were indispensable to the Roman ladies, and were frequently prized far more for their artistic merit than for the substance of which they were composed. Hair-pins constituted a very important article of the toilet, and were elaborately finished; the head usually represented figures delicately wrought. Mention is made of a hair-pin that cost £10,000. Among the relics of Pompeii and Herculaneum, now in the Royal Museum of Naples, is a pin that had belonged to the Empress Sabina; it represents the Goddess of Plenty, bearing in one hand the horn of Archeläus, and caressing a dolphin with the other. This pin is described by Winkelman in his letter on the antiquities of Herculaneum.

The necklace usually wound several times round the neck, the last circle falling on the bosom; the clasp was a magnificent cameo. We may judge of the delicacy of the workmanship, and of the beauty of the design, by the antique gems preserved in European collections.

Pearl bracelets of three or five strands, gold bracelets set with gems, loaded the arms and wrists of the Roman belles, rings encircled every finger, and rich girdles their waists. Many of these jewels have become historical. Thus Faustina's ring, we are told, cost £40,000; that of Domitia, £60,000; the bracelet of Cæsonia, £80,000; the ear-rings of Poppæa, £120,000; and those of Calpurnia, Cæsar's wife, twice that sum. The diadem of Sabina, as valuable for the workmanship as for the material, was estimated at £240,000.

The very garters of the Roman ladies were splendid trinkets, on which gold, silver, and precious stones were prodigally employed. Sabina, the younger, possessed a pair of garters valued at nearly £40,000, on account of the rich cameos that clasped them. The patrician dames, in their mad endeavours to rival each other in this species of ornament, spent a large part of their fortunes. The garters of those days were not used to fasten stockings with—the Romans wore no stockings—but a kind of drawers of fine linen. Sometimes the garter was worn on the naked leg, as bracelets are worn on the arms.

Nero offered to Jupiter Capitolinus the first cuttings of his beard in a golden vase enriched with very costly pearls.

Heliogabalus wore sandals adorned with precious stones of an inestimable price, and never wore the same pair twice.

Successive emperors vainly endeavoured, by various sumptuary laws, to arrest the progress of an extravagance that threatened ruin to all classes. Among other articles we find that gems were often made the subject of strict decrees.

Julius Cæsar, when he had attained the *apogee* of his fame and power, appears to have viewed with regret the relaxation of morals that had succeeded to ancient customs, for he made an edict prohibiting the use of purple and of pearls to all persons who were not of a certain rank; and even the latter were only permitted their use on days of public ceremony. Unmarried women were forbidden to wear precious stones, gems, or pearls. This proved the most severe blow that could be aimed at celibacy. Marriages increased to a surprising degree in every city of the empire. Rather than see themselves deprived of their ornaments, many women unhesitatingly perjured themselves.

The same edict forbade the use of litters, which was one of the fashions introduced from Asia.

The Emperor Leo, by a law of the year 460, which was the last of the sumptuary laws, made certain restrictions, which prove to what an extent his subjects carried extravagance. All persons, of whatsoever quality, were prohibited from enriching with pearls, emeralds, or hyacinths, their baldrics, and the bridles and saddles of their horses. All other gems were allowed for such purposes, but none whatsoever were permitted on the bit of the bridle. Men were allowed gold clasps to their tunics and mantles, and to exhaust all the resources of art in their workmanship; but it was forbidden to add thereunto any more precious ornaments.

In the dark ages that succeeded the ruin of the Roman empire, the productions and manufactures of the East sank in public estimation, and the Oriental trade, which had at one time threatened to exhaust the wealth of the West, ultimately dwindled into obscurity and utter insignificance.

CHAPTER IV.

GEMS AMONG THE GAULS, GOTHS, AND FRANKS.

During the Dark Ages.—Treasures found in Tolosa and Narbonne.—The Toledan Cross.—Trousseau of a Frankish Princess.—St. Eloi, the Patron of Jewellers. —Jewels of Charles the Great.—Heaven-guarded Jewels.

"They wear bracelets and armlets, and round their necks thick rings, all of gold, and costly finger-rings, and even golden corslets." —Diodor. Sic. b. v.

Though the Gauls delighted in bright-colored stuffs and golden ornaments, we find no record of their having worn precious stones. Their gems were merely hard stones, such as agates and jayet, of little value. Gold collars, armlets, and rings, adorned their warriors of renown; gold, silver, and coral, glittered on their arms; their woollen sagums, of which the modern blouse is the exact pattern, were checked with lively colors, or adorned with gold flowers and spangles; but they do not appear to have worn pearls or precious stones. This is the more surprising, as the Gauls delighted in everything brilliant, in everything that pleases the eye and amuses the fancy. To dazzle friends, and terrify enemies, was the great object of ambition; hence nothing could be more splendid, and, at the same time, more dreadful, than the aspect of a war-chief of that nation. That the gay and light-hearted Gauls, "who followed Cæsar under the standard of the lark, and marched singing to the capture of Rome, Delphi, and Jerusalem," should not have adorned themselves with the precious stones that must have fallen to their share; that the two Brenns, on their return to their druidical forests and hedge-fortified villages, should have left behind them this valuable portion of the rich spoils of the wealthiest cities of Greece and Italy, seems strange indeed.

When, however, the Gauls finally settled in the provinces they had overrun in devastating swarms,—when, submitting to the Roman yoke, the conquered had adopted many of the customs of the conqueror, they learned also to appreciate the beauty and value of gems.

The Goths, who delighted in pomp and magnificence, knew well the value of precious stones, and used them with profusion, to decorate their persons and their table gear. As early as the year 410, when the Eternal City fell into the hands of Alaric, we may judge of the value of the booty the Goths had gathered in their invasion by the nuptial gifts, which, in accordance with the custom of his country, Ataulphus, the brother of the conqueror, offered to the imperial bride. Fifty basins filled with precious stones of inestimable value, presented to the princess Placidia, formed no inconsiderable portion of the Gothic treasures, of which some extraordinary specimens might be quoted from the history of the successors of Alaric.

The descriptions that have come down to us of vases, bowls, and cups, famous for their beauty, their value, and the names of the personages who bestowed or received them, prove that precious stones were lavishly used. The Roman general, Ætius, presented to Thorismund, king of the Wisigoths, and successor of the great Alaric, a *missorium*, or gold dish, adorned with precious stones, and weighing five hundred pounds.

Theodric II., who, in 467, was deprived of his sceptre and his life by his brother Ewaric, had, while endeavouring to extend the boundaries of his dominions at the expense of the Romans, emulated their elegance and refinement.

The splendour and good order of the court of the Wisigothic prince presented, to a certain extent, the image of that of the emperors; he himself was only recognized as a barbarian "by the long locks that covered his ears."

The immense treasures in gems, and gold and silver vessels, amassed by the Ostrogoth and Wisigoth kings in Tolosa and Narbonne, bore witness to their taste in this respect. The first of these cities was considered the richest in Gaul. In 508, this capital of Ewaric opened its gates to the Franks, and the royal palace, with the treasures it contained, fell into the hands of Clovis; yet fame proclaimed that the greatest riches of the Wisigothic princes were not in Tolosa; the citadel of Carcassonne held, it was said, within its towers situated on inaccessible rocks, the deposit of the imperial spoils brought from captive Rome, by Alaric, during the preceding century; there, were to be seen the magnificent furniture of the temple of Solomon, king of the Jews, and the countless vases glittering with emeralds, that found their way to Rome after the sack of Jerusalem.

The immense number of curious ornaments of pure gold enriched with gems, found in the palace of the Wisigothic sovereigns, when Narbonne was pillaged, would scarcely be credited, were the details recorded by less trustworthy authors.

Among the rich ornaments of which Childebert, the son of Clovis, robbed the Church of Toledo, when he ravaged Spain, in 542, was a magnificent cross, enriched with wrought gems; which cross, so said the tradition, had belonged to King Solomon. Thirty chalices, fifteen patens, and twenty caskets in which the Scriptures were kept, all enriched with gems, were also carried off, and subsequently bestowed by the royal spoiler on various French churches.

As for the famous cross, he was so delighted with it, that he built a church in one of the suburbs of Paris, the plan of which he ordered should be cruciform, in honor of the Toledan cross, bestowing at the same time, that as well as other valuable ornaments on the new edifice.

As Christianity became more firmly established among Gauls, Goths, and Franks, the most valuable gems found their way into the treasuries of churches. After pillaging and destroying these sanctuaries of a God they ignored, the converted chieftains rebuilt and enriched them. They obeyed to the letter the injunction laid by St. Remy on the first Frankish prince who was baptized,—they burned what they had adored, and adored what they had burned.

Sometimes, indeed, they did both at once, as in the case of Rollo, the famous Norman chief, of whom it is recorded, that, on his death-bed, he was assailed by misgivings as to how he might fare in the next world, and as to whether he had acted wisely in forsaking the chance of the Walhalla of Oden for that of the Paradise of Christ. In this state of uncertainty he deemed it prudent to effect a compromise and secure a seat in both. To this end he caused a hundred Christian prisoners to be strangled, as a propitiatory offering to the infernal gods of his early creed, and bequeathed a hundred pounds weight of gold to the Christian churches.

Towards the close of the fifth century, the Franks, though in valor and headlong violence they had not degenerated from their ancient Teuton chiefs, had, however, renounced their voluntary poverty, and their systematic hatred of Roman civilization. Agathias calls the Franks the most civilized of barbarians. They were passionately fond of show, of costly dresses, of jewels, and of weapons enriched with the precious metals, and greatly favored all who traded in articles of luxury. Jews, Syrians, the inhabitants of Southern Gaul, and of other countries, were the dealers in such commodities; the Franks themselves never taking any part in commerce. Notwithstanding the innumerable perils, which, by land and sea, in times when every prince and noble was a chief of banditti or pirates, encompassed every step of the voyager laden with these coveted wares, the gains of the merchants were enormous.

The Franks differed only in dress and language from the Romans. As for their costume, it would have been esteemed rich, elegant, and picturesque in any age. Sidonius Apollinarius, the poet, courtier, and bishop, an ocular witness of what he relates, gives the following description of the brilliant appearance of the young chief Sigismer, when he entered Lyons to celebrate his nuptials with the daughter of one of the Burgundian chiefs. The royal youth was preceded and followed by horses whose caparisons sparkled with jewels . . . . "like the gold that gleamed on his dress, was his hair; beautiful in hue as the scarlet bands on his dress, was his complexion; his skin rivalled in fairness the milk-white silk in which he was attired. He advanced on foot, surrounded by the troop of chiefs of tribes (regulorum), and by a body of companions (antrustiones), terrible to behold, even in time of peace; they wore fur-boots, their legs were bare, and their short close-fitting sagums of green silk edged with purple, scarcely reached below the thigh. They bore swords suspended by rich baldrics, curved lances, axes of jayet, and shields of polished brass."

In the above quotation we find that the Franks, while they retained their own costume, had adopted some of the extravagant fashions of the Romans; they adorned even the harness and caparisons of their steeds with jewels.

When, in 584, Chilperic, king of the Franks, granted his daughter in marriage to Recarede, prince of the Spanish Goths, he gave her immense treasures, to which his queen Fredegunda added, with still greater liberality, a prodigious quantity of gold, silver, jewels, and costly garments. Fifty waggons were filled with the coffers containing the *trousseau*, plate, and ornaments of the

Princess Rigonthe. Having stopped, three leagues from Paris, to halt for the night, fifty men of her escort fled into the neighbouring domains of King Childebert, taking with them one hundred of the best horses with their *golden reins*, and two large chains of the same precious metal.

The passion of the Franks for display was never more conspicuous than in the early part of the seventh century, during the reign of King Dagobert, whose court rivalled in magnificence the pomp of Eastern monarchs. Precious stones sparkled in the golden girdles and fillets of the women and officers of the royal household; the monarch and his courtiers were clothed in robes of the rich China silk, brought from Asia at the risk of life by the Syrian merchants, and sold, literally, at its weight in gold. On solemn occasions, Dagobert occupied a throne of massive gold, wrought by no less a personage than the great Eligius, who, though subsequently bishop of Noyons, and the most popular saint in Gaul, was long the director of the royal mint, and the most skilful goldsmith-jeweller of the age.

Under Dagobert an immense number of gems were used to decorate the shrines and reliquaries of the saints, the crucifixes, crosses, church-vases, and other articles which Eligius (Saint Eloi) designed and executed for the king. The very dress of this artisan-bishop, during the life of the royal patron, with whom tradition and legend so intimately associate him, and before he himself had given up in his external appearance the vanities of this world, was exceedingly rich and elegant. The garments of Dagobert's favorite were thickly embroidered with gold and gems; he wore a golden girdle set with precious stones or pearls; his gowns were of fine linen embroidered with gold, and the border of his silken sagum was similarly adorned. The vandalism of modern revolutionists has melted and annihilated every specimen of the workmanship of this patron of goldsmith-jewellers; and though little more than half a century has elapsed since several interesting relics of the simple and severe style of that period were still extant, in the shape of church vases and clerical ornaments formerly belonging to St. Eloi, all have since been swept away by the unrespecting hands of republican iconoclasts.

The age of Charlemagne went far beyond that of Dagobert in prodigality towards churches; princes, bishops, and lords, vied with each other in the weight, the beauty of the workmanship, and the value of the gems of their offerings. Though during his lifetime this great monarch was very unostentatious and plain in his apparel, save on state occasions, his tomb contained a treasure in jewels and gold-plate, of which, unfortunately for the amateurs of art, but little remains. The canonization of Charlemagne, in 1166, furnished Frederic Barbarossa with a pretence for appropriating the golden chair, on which the newly-made saint was seated, clothed in his imperial robes, his jewel-covered sword by his side, his diadem on his head, his golden shield and gemmed sceptre suspended before him. Of these, and many other precious articles, the crown and sceptre alone remain; the one is in the imperial treasure-chamber of Vienna; the other, in the ancient treasury of the crown in Paris. Among the rich presents sent by Haroun-al-Rashid, to the Frankish sovereign, were precious stones of immense value.

The death of this great man seemed the signal for every misfortune to fall on the land. To internal divisions was added a more terrible foreign plague that had yet been known. The incursions of the Normans who, at their third invasion, in 845, laid siege to the capital itself, annihilating all commerce, and laying waste the adjacent country, had been, it was said, foreseen by the mighty mind of Charles; but his divided and weakened posterity was powerless to resist these northern pirates. Abbon, a contemporary who composed in barbarous Latin a poem on the siege of

Phyaris the Normans, reproaches the Franks with three chief vices, to which he attributes the disasters that afflicted the land. These vices are—pride, debauchery, and luxurious attire. In the description this writer gives of the dress of the Franks, we have another proof of their passion for jewels in the ninth century. "A golden brooch fastens the upper part of your garments; to guard your bodies from the cold, you cover them with Tyrian purple; your mantle must be a chlamyde loaded with gold; the belt that girds your loins is adorned with precious stones; even the shoes on your feet and the *cane* ye bear, must be covered with gold."

Under the successors of Charlemagne the clouds of barbarism began to gather, and the arts that ministered to piety or vanity declined rapidly. The West was fast retroceding to the savage state from which the civilizing genius of the mighty Karl had drawn it. The love of the beautiful in nature and art seemed entirely extinguished; brutal passions and degrading vices had usurped its place. Precious stones and precious metals seemed to have sunk back into their native mines; and it was not until the terrors inspired by the approach of the end of the world, predicted for the year 1000, again revived the piety, and consequently the liberality, of Christians, that we find frequent mention of gems.

The belief that the day of judgment, preceded by the advent of the Antichrist, was at hand, brought munificent gifts to the churches. Convinced that with the close of the century would end all necessity for wordly goods, people endowed basilicas and monasteries with their possessions. Fear, stronger than avarice, bestowed not only gold and jewels, but, in many instances, castles and wide domains. With the beginning of the eleventh century all apprehensions vanished, but the donations could not be retracted. That rapacity and avarice soon regained the ascendant, is evidenced by the fact, that Philip I., in the beginning of his reign, was influenced by his *prévôt* Etienne to lay violent hands on the treasures of the church of St. Germain-des-Prés. The gold, silver, and precious stones of the shrines, crucifixes, and vases, were about to become the booty of the prince and his evil genius, when, as the legend informs us, divine interposition prevented the sacrilege. The audacious provost, who had specially coveted the valuable cross brought from Spain by another royal robber, Chilperic, was struck with blindness as he extended his hand to grasp it. This miracle infused such terror into the heart of the king, that he withdrew, renouncing the unhallowed enterprise.

CHAPTER V.

THE TWELFTH, THIRTEENTH, FOURTEENTH, AND FIFTEENTH CENTURIES.

1100 to 1500.

Renaissance of the Jeweller's art.—Abbot Suger and Louis VI.—Religious Jewels in the ancient Treasure-House of St. Denis.—Louis VII.—Grace-cup of Thomas à Becket.—Jewels of a Royal Bride in the Twelfth Century.—Extravagance in Jewels during the Middle Ages.—Jewels lost at the Battle of Poitiers.—Stone-broidered robes of Philip the Bold, Isabella of Valois, and Richard II.—Jewels of Louis, Duke of Orleans.—Magnificence of the House of Burgundy.—Important part played by Jewels in the history of the Dukes of Burgundy.—Diamonds.

"Les François estoient bien peignés,
Les vis (visages) tendres et déliés;
Et si avoient barbes fourchées;
Bien dansoient en salles jonchées,
Et si chantoient comme sereines.
* * *
Grand coup (beaucoup) avoient de perleries,
Et de nouvelles broderies
Seulement le derroié (derrière)
Estoit de perles tout royé."—Old Author.

The Renaissance of Gems and of the jeweller's and goldsmith's art took place in the twelfth century, and met with great encouragement from the wise Suger, abbot of St. Denis and minister of Louis VI. This intelligent churchman liberally patronized an art in which he was himself a proficient. The superb pieces of plate bestowed by Suger and his king on the Abbey and Church of St. Denis, glittered with enamel and precious stones. So profusely were the offerings adorned with gems, that it was sometimes difficult to satisfy the requirements of the designers. Among the most costly gifts ordered by the minister were an altar screen and crucifix; on the latter six or seven workmen were alternately employed for two years, and finally it had well-nigh remained unfinished for want of gems, when two monks presented themselves offering for sale a quantity of beautiful stones, that had formerly adorned the table vases and cups of Henry I., king of England, and which had been given to different convents by Thibalt, count of Champagne. For the paltry sum of four hundred livres, Suger obtained jewels of immense value.

The crucifix on which these gems were employed is supposed to have been melted by the Leaguers in 1590.

The sanctuary and treasury of St. Denis formerly contained very great wealth in religious jewels, among which were the following specimens of the art of a still earlier period:—the altar-service and other apocryphal articles supposed to have been used by the patron saint, such as his ring and pastoral-staff covered with gold, enamel, and gems: the sceptre of Dagobert; the golden eagle set with sapphires and other gems, with which he clasped his mantle: the gifts of Charlemagne,—his *escrin*, or oratory, a small monument with three rows of arches inlaid with gold and gems, and surmounted by an antique cameo; his crown (scarcely authenticated) enriched with sapphires, rubies, and emeralds; his golden sceptre, six feet long; his sword, the hilt and scabbard of which were covered with gems; his gold spurs, etc., etc. There were also numerous shrines, crosses, and chalices of gold, enamelled and jewelled, due to the munificence of Charles the Bold, and the hanaper of oriental agate called Ptolemy's drinking-cup; this famous antique cameo, with its mounting, that dates from the ninth century, has been preserved to the present day.

Louis VII., following the example of his predecessor and the advice of Suger, equalled their liberality, and enriched the treasury of St. Denis with vases and shrines, studded with antiques richly set by the jewellers of his own day. Several of the gifts of this king are still extant, though specimens of the *orfèvrerie* and jewellery of the twelfth century are much rarer in France than in Germany and Italy, where good taste and piety caused them to be respected through every political change.

A precious relic of the gemmed orfèvrerie of the twelfth century is, we are told by a modern writer, still in existence in England. This is no less a curiosity than the grace-cup of Thomas à Becket. "The cup is of ivory mounted with silver, which is studded on the summit and base with pearls and precious stones. The inscription round the cup is, 'Vinum tuum bibe cum gaudio,' 'Drink thy cup with joy,' but round the lid, deeply engraved, is the restraining injunction, 'Sobrii estote,' with the initials T. B. interlaced with the mitre; the peculiarly low form of which stamps the antiquity of the whole*."

Gems were employed with like profusion in sacred and laical orfèvrerie; they enriched the vases of the Church and the table gear of the wealthy, as well as their personal ornaments. As a specimen of the quantity of jewellery supposed to be required by a regal bride in the thirteenth century, may be quoted the wedding gifts of Henry III. to his newly-wedded Eleonora of Provence, and which had cost him £30,000,—an enormous sum in those days. "Eleonora had no less than nine *guirlands*, or *chapelets* for her hair, formed of gold filagree and clusters of colored precious stones. For state occasions she had a great crown most glorious with gems, worth £1,500 at that era. Her girdles were worth 5,000 marks; and the coronation present given her by her sister, Queen Margaret of France, was a large silver peacock, whose train was set with sapphires and pearls and other precious stones wrought with silver. This elegant piece of jewellery was used as a reservoir for

* Miss Strickland.

sweet waters, which were forced out of its beak into a basin of silver chased*." Here we have another specimen of the skill of the French orfèvres of that age.

Not content with these nine garlands, the uxorious monarch added many more jewels to his queen's store. A farther sum of one hundred and forty-five pounds four shillings and fourpence was subsequently paid for "eleven rich garlands with emeralds, pearls, sapphires, and garnets†." In the list of Eleonora's valuables there is a royal crown set with rubies, emeralds, and great pearls; another with Indian pearls; and one great crown of gold ornamented with emeralds, sapphires of the East, rubies, and large oriental pearls‡.

All the courts of Europe carried extravagance in jewelry to an unlimited extent during the fourteenth, fifteenth, and sixteenth centuries. Whatever their personal embarrassments, their ruinous wars, the exhausted state of their exchequer, even though obliged to pawn one day the purchase made on the eve, sovereigns, princes, and nobles always seemed to find means to indulge this costly taste. The dress of the nobles during the Middle Ages was literally covered with gold and precious stones. At the defeat of Poitiers, the immense booty in coin, rich gold and silver plate, valuable jewels, gem-studded girdles, and trunks filled with costly apparel, so delighted the English and Gascons, that they treated their prisoners courteously, and not knowing what to do with them on account of their numbers, permitted them to return home on parole to collect their ransoms, which were to be paid, at Bordeaux, at Christmas. This the victors might safely do, for no man, however base, vile, and dishonest otherwise, would have dared to break his word in such a case; so infamous was such a dereliction held.

But the lords and knights, whose valour had been

* Miss Strickland. † *Idem.* ‡ *Idem.*

appreciated and respected by their enemies, met with a most ungracious reception at home. So universally were they hooted and insulted by their indignant countrymen, that they hardly dared show themselves in a town. The very peasants emulated the citizens in their angry reproaches: "Here be fair sons, truly," they cried, "who better love to wear pearls and precious stones on their hoods, rich goldsmith's work on their girdles, and ostrich feathers in their caps, than lances and swords in their hands. In such vanities and toys have they known how to spend the money we raised for the war; but to fall on the English, that knew they not."

The French, however, retrieved their losses during the succeeding reign, if we may judge by the value of the jewels and treasures of Charles V., seized at his death by the Duke of Anjou, and which was said to amount to the enormous sum of nineteen million livres.

In Wickliff's *Commentaries upon the Ten Commandments*, in the midst of a moral exhortation, he manages, in a few bold touches, to give a picture of the fashionable head-dress of his day: "And let each woman beware, that, neither by countenance, nor by array of body nor of head, she stir any to covet her to sin; not crooking (curling) her hair, neither laying it up on high, nor the head arrayed about with gold and precious stones; not seeking curious clothing nor of nice shape, showing herself to be seemly to fools: for all such arrays of women, St. Peter and St. Paul, by the Holy Ghost's teaching, openly forbid."

Colored stones and enamels were required for the heraldic and emblematic figures so much in fashion during the fourteenth century, that the very dresses of the ladies bore their coat of arms. The officiating garments of the clergymen were studded with gems. Isabella of France, the consort of King Edward II., sent to Pope John copes embroidered with large pearls. This was called apparel "broidered of stone," and the French embroiderers were held to excel in this sort of work. The following dresses, made for the Duke of Burgundy, Philip the Bold, on the occasion of his meeting with the Duke of Lancaster at Amiens, in 1391, give some idea of the use made of gems, in embroidery, as well as of the magnificence of the wearer. One was a surcoat of black velvet; on the left sleeve, which, according to the fashion of the day, hung as low as the hem of the dress, was embroidered a large rose branch, bearing twenty flowers; some of the roses were figured by sapphires surrounded by pearls, others by rubies; the buds were of pearls; the collar was similarly embroidered. The design round the button-holes was a wreath of Spanish genet, pearls and sapphires forming the pods. This was in honor of the ancient order of the genet, instituted by the kings of France. The body of the dress was embroidered with the Duke's initials:—P.Y.

The other dress was of crimson velvet, down each side of which was embroidered a large silver bear, whose collar, muzzle, and chain were of rubies and sapphires. A running pattern, in which were introduced the King's badge, the golden sun, and the Duke's initials, was embroidered round the border. With this robe the Duke wore a bracelet of gold set with rubies, sustaining a clasp, and chain also set with rubies. On these two dresses there was a weight in gold of thirty-one marks, and the making alone cost 2,977 livres.

When the elegant and unfortunate Richard II. was to be married to the little Isabella of Valois, preparations on a large scale were made for the nuptials in France and England. All the goldsmiths and embroiderers were set to work; their shops were filled with gold, silver, pearls, diamonds, and precious stuffs. The trousseau of the French princess was such a one as had never been given to any bride. Among her dresses was a "robe and a mantle, unequalled in England, made of velvet, embossed with birds of goldsmith's work, perched upon branches of pearls and emeralds." She had coronets, rings, necklaces, and clasps, to the amount of 500,000 crowns. The bridegroom was no less richly apparelled; he had one coat estimated at 30,000 marks.

The inventory of the effects of the Duke of Orleans makes one wonder how money could be found in those troubled times to purchase such a quantity of rich jewels as are there enumerated, and the settings of which were master-pieces of art. The liberality of this demi-monarch sometimes exceeded even his ample means, and he frequently borrowed on his gold plate to purchase new trinkets.

On every New Year's day, the Duke and his wife, Valentine de Milan, gave away jewels to an extravagant amount; necklaces, collars, reliquaries, rosaries, rings, girdles, and ear-rings set with gems, were profusely lavished. The churches and saints came in for their share. In 1392, the Duke presented to the shrine of *Monseigneur* Saint Denis, a clasp of gold garnished with three sapphires and three large pearls, with a ruby in the centre. The Duke bought but to give away. A departure, a return, a marriage, a christening, every event in the lives of those who surrounded him, was a motive for the bestowal of a jewel. He made presents to the king himself, to the queen, the dauphin, the princesses. Not a bishop was consecrated but the Duke sent him a piece of plate; while his gifts to his royal relations always consisted in jewels. In 1395, he sends to the pope "a jewel of gold in the shape of a head of *Madame* Sainte Catherine, supported by two golden angels," and adorned with balass rubies, sapphires, and large pearls.

The catalogue of table ornaments and other gold and silver plate, enamelled and inlaid with gems, of this prince, shows the degree of excellence which the workmen of that age had attained, and the prodigal liberality with which he encouraged them.

It is to be lamented that so very few of these jewels, the setting of which was so delicate, rare, and curious, have come down to us. This, however, in the case of the Duke of Orleans, is easily accounted for; the greater part of his beautiful trinkets, wrought at great expense, by the best workmen of the day, were, after his death, sold by weight to the Lombard changers, who melted the gold, and carried that and the gems out of the kingdom.

But, to find the luxurious splendour of the middle-ages in all its éclat, we must turn to the puissant House of Burgundy, from Philip the Bold, to Charles the Bold; these great Dukes, who tributed to the beautiful in art a species of worship, and whose brilliant court outshone that of their feudal lieges, the kings of France, and cast into the shade the rude grandeur of the German emperors, had magnificent collections of jewels, as well as of gold and silver plate of exquisite workmanship. Philip the Bold and his son, John the Fearless, spent largely in such articles; Philip the Good and Charles the Bold made it a point of honor to spend ten times more in the same manner. It is doubtful whether any sovereign in Europe made purchases as numerous and costly as those which absorbed the revenues of the House of Burgundy—purchases made too with rare taste and excellent judgment. Not only their own jewellers, but those of Florence, Lucca, Genoa, and Venice, and also the money-changers, who played the parts of usurers and lenders on securities, were constantly bringing them some marvel in the shape of a jewel, or a vase. Gems that can be traced back to the last of these mighty lords retain their fame to this day, and are among the most valued possessions of the Crowns of France, Austria, and Tuscany.

Although the art of cutting and polishing the diamond has been erroneously attributed to Robert de Berquen, who flourished in the reign of the last Duke, Charles the

Bold, diamonds were in such esteem during the reigns of his father, grandfather, and ancestors, that the conviction is forced upon us, that the art was well known before, though De Berquen probably perfected it. Diamonds form a chief item on all festive occasions, and are mentioned in every page of their history, whether the narrative be of triumph or defeat, marriage or death.

When the son of Philip the Bold was married at Cambray to the Princess of Bavaria, in 1395, the Duke distributed splendid *diamonds* to the ladies who were present. His gifts were estimated at 77,800 francs. The ladies were dressed in cloths of gold and of silver, brought from Cyprus and Lombardy. At the tournament which followed the great banquet—served by the great officers of the crown, mounted on their *chevaux de parade*, the Duchess of Burgundy presented to the victor the clasp or *fermail* of diamonds she wore on her bosom.

When the Duke, who concluded in the following year, the treaty of marriage between Richard II. and the Princess Isabella of Valois, met the English monarch in Calais, presents were exchanged between them. That of Richard was a fine diamond; Philip, who was never outdone in generosity, presented the king with two pieces of gold plate, representing the Passion, and the Saviour in the Sepulchre, and also with a piece of damask richly embroidered in gold.

On the morning of the 27th of October, of the same year, the Dukes of Lancaster and Gloster, and the Earl of Rutland, waited on the French king, to receive his commands with regard to the ceremonies to be observed, and the dresses to be worn by the two kings at the appointed meeting. Charles received them very graciously, and gave to each a fine diamond. The Dukes of Berry, Burgundy, and Bourbon, having gone to Richard on a similar message, he replied, that peace and friendship were not proved by magnificent robes, and that it needed not so many ceremonies in an interview so entirely friendly and cordial.

When the Duke of Burgundy, in whose hands the regency of France had been placed during the insanity of the king, undertook to prevent the widow of John, Duke of Brittany, from taking her children to England, he exercised his munificence to good purpose. The largesses of the Duke proved more powerful than force of arms, in counteracting the policy of the English prince, who gained the widow for his bride, but not the possession of the duchy in the person of its young heir. The victory must have cost the regent no small sum, the presents consisting, as was the custom, in rich jewels. At the conclusion of the banquet offered to him by the Duchess, in Nantes, whither he had gone on this political errand, in 1402, Duke Philip presented her with a rich crown of *crystal*, and another of gold, set with pearls and precious stones. He gave the young Duke a clasp of gold, adorned with rubies and pearls, a beautiful diamond, and a quantity of silver plate. To each of the younger boys, Arthur, Earl of Richmond, and Jules, Count of Bretagne, he gave a gold collar set with pearls and rubies. The Countess of Rohan, aunt of the Duchess, accepted a very fine diamond, and to each of the ladies and damsels present the gallant negociator presented a rich clasp. The lords in waiting, and the officers of the Duchess' household, came in for an ample share in this princely distribution. The result could scarce be doubted. The Duke, who certainly had deserved to win golden opinions, was appointed guardian of the children, and of their inheritance. At an entertainment he gave to the king and the court, at the Louvre, in 1403, the Duke of Burgundy presented to his guests, gifts, among which figured eleven diamonds, valued at 785 crowns.

At the marriage of his second son, he gave to all the lords of the Low Countries who were present, robes of green velvet and white satin, and jewels to the amount of ten thousand crowns.

Two years after his visit to Brittany, this high and mighty founder of the House of Burgundy, whose immense revenues constituted him one of the wealthiest of European princes, died a bankrupt; all his store of rich goods, his collection of magnificent jewels, would not have sufficed to pay his debts, unless some portion of his territorial possessions had been disposed of. To maintain undivided the splendid inheritance of her husband, to his children, the strong-hearted Duchess went through the degrading ceremony of declaring her husband a bankrupt. (*For this ceremony, see chapter on girdles.*)

At the entrée of Louis the Eleventh into Paris, at his accession in 1406, Philip the Good, Duke of Burgundy, as usual, far excelled all other nobles in magnificence. His saddle and chanfrin were adorned with diamonds. His apparel sparkled with diamonds. The purse that hung at his belt, was the subject of universal admiration; it seemed one mass of gems. The jewels he wore were valued at one million francs.

Amid the crowd of princes and lords that witnessed the inauguration of the new reign, none held such state as Duke Philip. When he visited the churches he had never less than a hundred knights in his suite, many of whom were princes and powerful lords. His archers were splendidly equipped. He changed his jewels daily; sometimes his belt was of diamonds, or his rosary of precious stones; at other times, his bonnet would be entirely covered with gems. The Parisians, who had seen so many princes that they now never troubled themselves to look at them, ran tumultuously through the streets to catch a glimpse of the Duke of Burgundy, whenever he was expected to pass.

Philip the Good, grandson of the insolvent Philip the Bold, was, at his own death, in 1467, the wealthiest prince of his age, although in liberality he had exceeded his predecessors. He left four hundred thousand crowns in gold, seven thousand two hundred marks in silver plate, and an immense value in rich hangings, jewels, gold plate adorned with precious stones, besides a fine library; his furniture alone was valued at two millions in gold. No sovereign in Europe surpassed in power this "Great Duke of the West," under whom were united all the provinces of the Low Countries from the Ems to the Somme—a union that had given a new impulse to industry, commerce, and the fine arts, already so flourishing in those regions. The arts that ministered to the requirements of luxury, had been carried, during this last reign, to an extraordinary degree of perfection. The magnificence of the costumes, arms, jewels, and furniture, exceeded anything ever seen before—one might almost add, or since. The fifteenth century, denominated the age of iron, on account of the beauty and finish of the armours, and other steel articles, might with equal reason be called the age of gold and gems. The art of the jeweller, depressed in France by the indifference as well as by the sumptuary laws of Louis XI., was developed in the states of Burgundy and Flanders to a surprising degree of refinement and elegance. Velvets, satins, cloth of gold and cloth of silver,—every costly material, was still farther enriched by the addition of gold and gems. Marvels of art were produced by the *orfèvres* that preceded Benvenuto Cellini, and the exquisite workmanship frequently exceeded in value the gold and gems used so profusely.

Charles, surnamed by English writers the Bold, but by the French more appropriately designated as the Rash,—Charles, with whom all the wealth, splendour, and power of the House of Burgundy was to fall, surpassed all the princes of his line in pomp. When at Aix-la-Chapelle, in 1473, the magnificence of his church-plate, displayed in the Chapel arranged for him, was the admiration and wonder of the simple Germans. Four tables covered with cloth of gold displayed immense wealth. Among other articles were the twelve Apostles in silver gilt; ten figures of saints in massive gold; a quantity of large gold and silver crucifixes of the most exquisite designs and enriched with diamonds; four great massive silver candelabras, and two of massive gold; a shrine of gold covered with diamonds, containing relics of Saint Peter and Saint Paul; a tabernacle entirely of gold. The most precious of all the rare and beautiful things there, was a lily of diamonds, containing a nail and a bit of the wood of the true cross, over which was set a diamond, "two fingers' breadth in length."

At the interview between the Duke of Burgundy and the Emperor at Treves, a few days after, the Duke presented himself, fully equipped in splendid armour, over which he wore a mantle trimmed with gold and diamonds to the amount of 200,000 ducats. The Emperor wore a long robe of cloth of gold embroidered with pearls.

Wherever the Duke made a ceremonious entrée, he appeared with like magnificence. When he entered Dijon this same year, he was resplendent with pearls and diamonds. At his triumphant entrance into Nancy, in 1475, the ducal crown that he wore over his crimson bonnet was so rich with diamonds and pearls, "it was worth a whole duchy."

When this fool-hardy prince was defeated at Granson, in 1476, his magnificent plate was mistaken by the rude and ignorant victors for pewter, and sold for a few deniers, they pronounced the superb gold and vermeille-plate; cumbersome litter, and sold it as copper for a mere trifle. The magnificent hangings of silk and velvet, broidered with pearls; the gold cords which fastened the duke's pavilion; the cloth of gold-damask; the Flanders lace; the carpets and hangings of the famous Arras manufactory, of which an immense quantity was found in the coffers,—all were cut up, and distributed by the yard, like common stuffs in a village shop. The Duke's tent was surrounded by five hundred others, in which were lodged the lords of his court, and the officers and dependents of his household. The ducal tent was distinguished on the outside by the scutcheon, enriched with pearls and precious stones; within, it was hung with crimson velvet embroidered with gold foliage and pearls; apertures had been arranged for windows, the glass panes of which were encased in gold. The chair in which he sat, when he received ambassadors, and gave solemn audiences, was of massive gold. His suits of armour, swords, daggers, ivory-mounted lances, were marvellously wrought, and the hilts were covered with rubies, sapphires, and emeralds. His seal, which weighed two marks in gold, his velvet-bound tablets, in which were his portrait and that of his father, his collar of the toison, and an infinite number of precious jewels, were plundered, divided, and destroyed.

The tent which served as a chapel, contained an enormous amount of wealth; some of the articles in it have been already mentioned as exciting the admiration of the inhabitants of Aix-la-Chapelle, two years before. The treasure of the Duke fell also in the hands of the confederated Swiss. It was so great that instead of counting, or weighing the coin, they distributed it among them by hatfuls. The history of the three famous diamonds, and of the other chief gems of the crown of Burgundy, lost on that terrible day, is given in Chapter I., Part Third.

The Spaniards and Italians of the fourteenth and fifteenth centuries, displayed the most unbounded luxury in apparel. Silken tissues, cloth of gold and silver, loaded with embroidery and gems, are seen in all the portraits of that day. The dress of a young Italian duchess is rich beyond measure. In the pictures representing the Queen of Cyprus, surrounded by noble Venetian dames, the corsages of the ladies are thickly studded with gems, so tastefully arranged however, that, notwithstanding their profusion, the effect is not heavy.

The coiffure of Beatrix of Este is extremely rich and beautiful. It consists of pearls, and of an ornament of two precious stones and large pendant pearls, placed near the ear. A string of large pearls encircles the throat and bosom. Eleonora, Princess of Portugal, and Frederic, Duke of Urbino, are covered with gems.

Among the Italian princes most distinguished for valor, taste for the fine arts, and magnificence, was Martino II., Lord of Verona, Brescia, Parma, and Lucca, who died in 1351. His military costumes and armour were the richest of that century. The House of Visconti was also famed for its splendour. On the occasion of the coronation of Galeas Visconti, the ducal coronet of gems and pearls placed on his head by the Imperial plenipotentiary was estimated at 200,000 florins (416,850 francs, or £16,666).

The costume of a Milanese noble consisted in a bonnet of black velvet, surrounded by a coronet of pearls; a dress of gold brocade, edged with crimson velvet embroidered with pearls; a rich pearl collar with a jewelled clasp and wrought gold-hilted sword, completed this costume, of which a picture by Bartholomew Montagna, of the year 1498, gives an exact representation.

In magnificence the Spaniards kept pace with the princes of France and Italy, while in lavish generosity they excelled all save the princes of the House of Burgundy. Spain was the first to introduce the Oriental fashion of presenting jewels to the guests at a banquet. Among innumerable instances of this regal hospitality, we will only quote that of the Count de Haro, who, in 1440, having had the honor to receive the Queen of Navarre and her daughter in his domains, gave fêtes, the description of which seems taken from some fairy tale. At the close of an entertainment, the noble host, approaching the Princess, knelt before her, and presenting her with a jewel of great price, thanked her for the honor she had done his house. The same ceremony was repeated to the Queen; and to every lady present a rich jewel was also given, not one being forgotten; each had a diamond, emerald, or ruby ring. Every knight and gentleman in the royal cortège had a gift of a fine mule, or a piece of costly silk or brocade.

When the sovereigns of Castile and France met on the banks of the Bidassoa, in 1464, to hold a conference, they exhibited a striking contrast, not only in their own persons, but in those of their followers. The avaricious and crafty Louis XI., dressed in the mean, coarse attire which he usually wore, was followed by a suite who, imitating with courtier-like servility their king, had adopted a similar costume; while Henry IV. appareled in the rich and becoming Spanish garb of the day, covered with magnificent jewels, was attended by his Moorish guard, splendidly equipped, and by a train of nobles whose dress and equipage was of the most gorgeous and sumptuous description. Don Beltran de la Cueva, the favorite of the Castilian monarch, distinguished himself by the splendour of his jewel-studded dress, his very boots being embroidered with pearls. His barge was decked with cloth of gold, and the sails were of brocade.

Nearly all the European trade of precious stones, at the close of the fifteenth and commencement of the sixteenth centuries, was in the hands of the Fuggers and Obwexers, rich merchants of Augsburg.

CHAPTER VI.

THE SIXTEENTH CENTURY.

1500 to 1560.

Louis XI.—Charles VIII.—Francis I.—Cardinal d'Amboise's Encouragement to Arts.—Jewels of that day.—Anecdote of the Jewels of the Countess of Chateaubriand.—Henry II. of France.—Henry VIII. of England.—Catharine of Arragon.—Anne of Cleves.—Elizabeth.—Passion of that Queen for Jewels.—Jewels of Mary, Queen of Scots.

" To-day the French
All clinquant, all in gold like heathen gods,
Show down the English; and to-morrow they
Made Britain, India; every man that stood
Show'd like a mine. Their dwarfish pages were
As cherubins, all gilt: the madams, too,
Not us'd to toil, did almost sweat to bear
The pride upon them."

Louis XI., who in his own person exhibited the most sordid avarice, whose meals were not unfrequently served in pewter dishes, whose greatest outlay in jewelry or orfévrerie consisted in the little *enseignes*, or images of saints, with which he decorated his mean old hat*—even many of these were but leaden medals—could not be expected to patronize the art of the jeweller; though, when

* If any farther proof was wanting of the parsimonious personal habits which history attributes to this monarch, it has been furnished by two very curious parchments lately sold at the public auction rooms in the Rue Drouol. One was a receipt given to the treasurer of Louis XI. by the King's tailor for the sum of 30 sols, for putting a new pair of sleeves to an old leather pourpoint of His Majesty, and the other a receipt from the royal shoemaker for 15 deniers, for furnishing a box of grease for the King's boots. Gems would have proved an adjunct as superfluous to his toilet, as in the case of the unclean beast, before which pearls are thrown.

stimulated by fear or covetousness, he sometimes endeavoured to propitiate heaven by rich presents of plate to the churches.

After the short reign of Charles VIII., which can only be said to have commenced in 1491, when he took upon himself to govern, and which ended in 1496, we have that of Louis XII. The court of this monarch was not only more magnificent, but also far more refined and elegant than that of his predecessors. The dawn, precursor of the brilliant sun that was to render the reign of Francis I. so glorious, was hastened by the judicious encouragement the arts received from Cardinal d'Amboise. The munificent patronage of the minister, while it fostered genius, and gave it the powerful impulse that led to more ample development at a later period, also introduced a new feature in the arts of the jeweller and goldsmith. So good a judge could not but appreciate the Italian style; with the quantities of exquisitely-wrought articles he imported from Milan and Genoa, with the celebrated artists he invited into France, was introduced the Italian character, which, surviving the age of Louis XII., stamped that of his successor. Of the value and quantity of the jewels and plate collected by the Cardinal, some idea may be had from the fact, that at his death, he left to one of his nephews a beautiful vase, estimated at 200,000 crowns, all his silver-gilt plate and a part of his silver plate to the amount of 5,000 marks; and yet all these valuables were apart from his pontifical inheritance, which remained untouched, and was estimated at two millions; or the furniture of his Castle of Gaillon, which he left to another nephew.

The portraits we have of Francis I., and of the lords and ladies of his court, sufficiently betoken the taste of that period. Men and women are covered with jewels; girdles, baldrics, coifs, gold chains, necklaces, and rings, studded with precious stones, were worn by both sexes. Well might one of their contemporaries exclaim, "Those people carry their fields and mills upon their backs."

The rich ornaments worn during the reign of Francis I., and his immediate successors, were as valuable for the exquisite setting as for the gems of which they were composed; and this is not surprising, when such men as Leonardo da Vinci, Rosso, Nicolo Primaticio, and their pupils, did not disdain to furnish the designs for these superb jewels. In the inventory of those of Henry II., made in 1560,—of the rings, ear-rings, bracelets, and medallions, many are mentioned as having been wrought by Benvenuto Cellini. Unfortunately all have long since disappeared. This artist excelled in making the medallions then called *pourtraits* or *enseignes* of gold, which men wore in their hats and women in their head-dresses.

In 1538, Benedict Ramel (Ramelli) had already wrought after this fashion a portrait of the king that cost 300 *livres tournois*. In the reign of Henry II. these enseignes, such as they are described in the inventory, were marvels of the jeweller's art; enamel, gold, silver, gems and fine stones, were combined in the most delicate and ingenious manner. The following extracts from the inventory will give some idea of this ornament:—"A golden enseigne, representing several figures, garnished all around with small rose-diamonds; an enseigne of gold, the ground of lapis-lazuli, the figure representing a Lucretia; an enseigne with a gold setting, the figure being a Ceres on an agate, the body of silver, the dress of gold; an enseigne of a David and a Goliath, the head, arms, and legs of agate."

Brantome gives a description of the dresses of the ladies who represented nymphs and goddesses in a show with which the Queen of Hungary entertained her imperial and royal relatives, the Emperor Charles V., his son, the King of Spain, and Queen Eleonora. The six Oreades each wore a diamond crescent on her brow. Pales, the goddess of shepherds, and her six nymphs, were dressed in cloth of silver covered with pearls. The goddess Pomona, a child nine years old, daughter of one of Eleonora's Spanish ladies, wore a head-dress composed of emeralds to represent the fruit over which she is supposed to preside. To the Emperor and his son she presented palm branches of green enamel, loaded with large pearls and precious stones; "rich to the view, and of inestimable value:" to Queen Eleonora the diminutive deity's offering was a fan, in the centre of which was a mirror all garnished with gems of exceeding richness.

In a show of the same kind prepared by the city of Lyons in honor of the visit of Henry II., the goddess Diana and her nymphs wore boots of crimson satin; large cords of rich pearls were wreathed in their tresses, which were also adorned with quantities of precious stones of great value: on their brows they wore a silver crescent blazing with small brilliants.

It was in the reign of Francis I. that devices, as they were called, became so fashionable, that almost every personal ornament was made to express the state of mind of the wearer, the donor, or the recipient. Sometimes indeed these high-flown and ornate conceits were so elaborate, and far-fetched, that in lieu of conveying a sentiment or ruling passion, they became very puzzling enigmas. The noblest personages of the court exercised their wit and ingenuity in the invention of these mottoes. The following rather ungallant action of the gallant knight Francis I., related by Brantome, shows the importance attached to these fanciful allusions. Mademoiselle de Helly, afterwards Duchess d'Estampes, having supplanted the Countess of Chateaubriand in the affections of Francis I., urged her royal admirer to claim all the rich jewels he had bestowed on his former mistress, not indeed on account of their metallic value, but for the sake of the beautiful devices thereon engraved, or otherwise signified; the which devices had been composed by the king's sister, Margaret of Navarre, she being a great mistress in the art. The King having, in pursuance of the ungenerous behest of his fair tempter, sent a gentleman to demand the jewels of the Countess, she, feigning illness, bade him come again in three days and she would have them ready. In the interim, sending for a goldsmith, the angry lady caused all the beautiful tokens of her perjured lover to be melted, without regard for the quaint and witty devices they bore. On the messenger's return, she presented them to him converted into golden ingots. "Go," said she, "bear these to the King, and tell him I return him the material of that which he so liberally bestowed upon me. As for the devices, I had so deeply impressed them in my heart, and hold them so dear, I could not permit that any other person should possess them, or take the like pleasure in them." When the message and gifts in their altered form were brought to Francis, he sent the ingots back, bidding her be told, he had not desired them for their value, for he had purposed compensating that richly, but for the sake of the devices; since those were destroyed, he would not have the gold: adding "that she had shown more boldness and great-heartedness than he had thought was in woman."

The *Comptes Royaux* bear witness to the taste of Francis for orfévrerie. Among a number of purchases is a gold belt, garnished with precious stones, a gold border garnished with rubies and diamonds, and a collar garnished with diamonds, all of which were purchased in 1535, of Robert Rousset, orfévre, of Paris, for the sum of 3,600 *livres tournois*, or £142 4s. 3d.

From the reign of Francis I. to that of Louis XIII. the majority of the jewels worn were set with pearls and colored gems. Sometimes a diamond was placed in the centre of a clasp of precious stones. Pearls continued to be worn in preference to all other ornaments until the death of Maria The esa, of Austria, when brilliants came into fashion.

The magnificence displayed in gems at the famous meeting of the cloth of gold, was in accordance with the extravagance exhibited in other things. The banqueting chamber in which Henry VIII. entertained Francis I., was hung with tissue, raised with silver, and framed with cloth of silver, raised with gold. The seams of the same were covered with broad wreaths of goldsmith's work, full of precious stones and pearls. In this chamber was a cupboard seven stages high, all plate of gold, and no gilt plate*. The foot carpet of the English queen's throne was embroidered with pearls. When Charles V. departed from Calais for Gravelines, his aunt gave him a beautiful English horse, with a footcloth of gold tissue, broidered with precious stones.

The scarf, or Spanish mantilla, worn by Catharine of Aragon at her marriage, had a border of gold, pearls, and precious stones, five inches and a-half broad, and so long that it veiled great part of her visage and person. This scarf must have been of itself a very great weight for a delicate young woman.

The King of England appears to have been as fond of ornaments as his French cousin, if we may judge by his pictures, and the description given of some of his suits. When he met his bride, Anne of Cleves, he was appareled in a coat of purple velvet, made somewhat like a frock, all over embroidered with flat gold of damask, with small lace mixed between, traverse-wise, so that little of the ground appeared . . . . the sleeves and breast were cut and lined with cloth of gold, and clasped with great buttons of diamonds, rubies, and oriental pearls; his sword and girdle adorned with stones and special emeralds; his cap garnished with stones; but his bonnet was so rich of jewels, that few men could value them . . . . Besides all this, he wore a collar of such balas, rubies, and pearls, that few men ever saw the like. Henry's wedding coat of crimson

* Miss Strickland's *Queens of England*.

satin, slashed and embroidered, was clasped with great diamonds.

Anne of Cleves' wedding dress was "a gown of rich cloth of gold, embroidered very thickly with great flowers of large orient pearls." She wore a coronet of costly gems, and jewels of great price about her neck and waist. The partlet, a sort of habit-shirt worn by ladies, was usually embroidered with silk and gold thread; but those of women of high rank were covered with gems.

The dress of Queen Mary, when she was married to Philip of Spain, was in the French style,—a robe richly brocaded on a gold ground, with a long train splendidly bordered with pearls and diamonds of great size. The large rebras sleeves were turned up with clusters of gold, set with pearls and diamonds; her chaperon or coif was bordered with two rows of large diamonds. The close gown, or kirtle, worn beneath the robe, was of white satin wrought with silver. On her breast, the Queen wore that remarkable diamond, of inestimable value, sent to her from Spain by her royal bridegroom, through the hands of the Marquis de Las Naves.

The rich white satin dress worn by Elizabeth at a tournament given in Mary's reign, on the 29th of December, 1554, is described as having been passamented all over with large pearls.

No European sovereign ever manifested so inordinate a passion for ornaments as did Elizabeth. Though everything that came as a gift was welcomed, jewels were more acceptable than any other offering. The crowd of obsequious courtiers over whom she held sway more tyrannically than ever did eastern despot over his subjects, were constantly impoverishing their coffers to minister to this foible.

In fact, such was the Queen's passion for jewels, and to such an extent was it gratified, that from the description left us of those she wore, and of the value and number of those presented to her, it would seem as though few could have been left to the English nobles who tributed to her so slavish an homage; yet all were bedizened with gems.

Queen Elizabeth employed precious stones profusely for other purposes besides the adorning of her person, if we are to believe a contemporary poet, who probably saw what he described, and expressed his admiration of her equipage, on the occasion of her visit to Tilbury, in the following lines:—

" He happy was that could but see her coach,
The sides whereof beset with emeralds
And diamonds, with sparkling rubies red,
In checkerwise, by strange invention,
With curious knots embroidered with gold."

The regal mantle and train of Mary, Queen of Scots, on the occasion of her nuptials with the Dauphin of France, was of a blueish grey cut velvet. It was of marvellous length, full six toises, covered with precious stones, and was supported by young ladies. The weight must have been very great, and yet the bride danced with this train twelve yards long, borne after her by a gentleman following in the devious maze!

The jewels of Mary, Queen of Scots, were numerous and very fine. They have, moreover, acquired great historical celebrity from the frequency with which they w ee claimed by their unfortunate mistress, in her appeals for mercy and justice during her long captivity, and the rapacity with which her royal jailor, and other enemies, sought or retained the possession of these glittering spoils.

The iniquitous manner in which Elizabeth possessed herself of her captive's property, is too interesting to be omitted here.

" If anything farther than the letters of Drury and Throgmorton be required to prove the confederacy between the English Government and the Earl of Moray, it will only be necessary to expose the disgraceful fact of the traffic of

Queen Mary's costly parure of pearls, her own personal property, which she had brought with her from France. A few days before she effected her escape from Lochleven Castle, the righteous Regent sent these, with a choice collection of her jewels, very secretly to London, by his trusty agent, Sir Nicholas Elphinstone, who undertook to negociate their sale, with the assistance of Throgmorton, to whom he was directed for that purpose. As these pearls were considered the most magnificent in Europe, Queen Elizabeth was complimented with the first offer of them. 'She saw them yesterday, May 2nd,' writes Bodutel La Forrest, the French ambassador at the court of England, 'in the presence of the earls of Pembroke and Leicester, and pronounced them to be of unparalleled beauty.' He thus describes them:—'There are six cordons of large pearls, strung as paternosters; but there are five-and-twenty separate from the rest, much finer and larger than those which are strung; these are for the most part like black muscades; they had not been here more than three days, when they were appraised by various merchants; this Queen wishing to have them at the sum named by the jeweller, who could have made his profit by selling them again. They were at first shown to three or four working jewellers and lapidaries, by whom they were estimated at three thousand pounds sterling, (about ten thousand crowns,) and who offered to give that sum for them. Several Italian merchants came after them, who valued them at twelve thousand crowns, which is the price, as I am told, this Queen Elizabeth will take them at. There is a Genoese who saw them after the others, and said they were worth sixteen thousand crowns, but I think they will allow her to have them for twelve thousand.' 'In the meantime,' continues he, in his letter to Catherine of Medicis, 'I have not delayed giving your Majesty timely notice of what was going on, though I doubt she will not allow them to escape her. The rest of the jewels are not near so valuable as the pearls. The only thing I have heard particularly described, is a piece of unicorn richly carved and decorated.' Mary's royal mother-in-law of France, no whit more scrupulous than her good cousin of England, was eager to compete with the latter for the purchase of the pearls, knowing that they were worth nearly double the sum at which they had been valued in London. Some of them she had herself presented to Mary, and especially wished to recover; but the ambassador wrote to her in reply, that 'he had found it impossible to accomplish her desire of obtaining the Queen of Scots' pearls, for, as he had told her from the first, they were intended for the gratification of the Queen of England, who had been allowed to purchase them at her own price, and they were now in her hands*.'

Inadequate though the sum for which her pearls were sold was to their real value, it assisted to turn the scale against their real owner.

In one of her letters to Elizabeth, supplicating her to procure some amelioration of the rigorous confinement of her captive friends, Mary alludes to her stolen jewels: "I beg also,' says she, 'that you will prohibit the sale of the rest of my jewels, which the rebels have ordered in their Parliament, for you have promised that nothing should be done in it to my prejudice. I should be very glad if they were in safer custody, for they are not meat proper for traitors. Between you and me it would make little difference, and I should be rejoiced if any of them happened to be to your taste, that you would accept them from me as offerings of my good-will.'

"From this frank offer it is apparent that Mary was not aware of the base part Elizabeth had acted, in purchasing her magnificent parure of pearls of Moray, for a third part of their value. The parsimonious English sovereign had, after all, been too precipitate in concluding her bar-

* Miss Strickland.

gain with Moray's agent, Sir Nicholas Elphinstone, since she might have saved her 12,000 crowns, and had the pearls for nothing from the rightful owner*."

Poor Mary perseveringly, but vainly, endeavoured, throughout the long term of her imprisonment, to shame or threaten her spoilers into a restitution of her jewels. To shame they were impervious, and as for her threats, they knew she was in no condition to fulfil them. Those to whom she had shown most love and favor were those most eager to appropriate a share of her spoils. Lady Moray, whom she had so liberally remembered in her will when she had thought herself in danger of dying, repaid her generosity by helping herself largely from her helpless mistress' store of jewels, as the following extract from one of Queen Mary's letters shows :—

"Since the which, we are informed, ye have tane in possession certain of our jewels, such as our H of dyamont and ruby, with a number of other dyamonts, ruby, perls, and gold work, whereof we have the memoir to lay to your charge. Which jewels, incontinent after the sight thereof, ye shall deliver to our right trusty cousins and counsellors, the Earl of Huntley, our lieutenant, and my Lord Setoun, who will, on so doing, give you discharge of the same in our name, and will move us to have the more pity of you and your children. Otherwise, we assure you, ye shall neither bruick (enjoy) lands nor goods in that realm, but to have our indignation as deserves. Thus wishing you to weigh with good conscience, we commit you to God.

"MARIE R.

"From Futburg, the 28th day of March, 1570."

"The great H of diamonds and rubies particularly demanded by Mary, was an ornament for the breast in that form, called 'the Great Harry,' having been originally

* Miss Strickland.

given by Henry VII. to his daughter Margaret, on her marriage to James IV., as part of her rich bridal outfit, so that it really formed no part of the crown-jewels of Scotland, but was Mary's private property; she had a peculiar value for this Tudor heir-loom*."

* Miss Strickland.

CHAPTER VII.

THE SIXTEENTH AND SEVENTEENTH CENTURIES.

Sumptuary Edicts of Charles IX. and of Henry III.—Dress of Henry IV.—Dress of Mary de Medicis.—A costly Kerchief.—Extravagance of the Court during minority of Louis XIII.—Magnificence of the Court of Louis XIV.—The fatal Bouquet.—Curious Letter of James I.—Jewels worn in Hats.—An expensive Suit of Clothes.

"... . Wilt thou ride, thy horses shall be trapp'd,
Their harness studded all with gold and pearl "
TAMING OF THE SHREW. INDUC. 2.

"MARGARET. . . . I saw the Duchess of Milan's gown that they praise so.
HERO.—O that exceeds they say.
MARGARET.—By my troth, it's but a night-gown in respect of yours; cloth o' gold, and cuts, and laced with silver, set with pearls, down sleeves, side sleeves and skirts round, under-borne with a bluish tinsel; but for a fine, quaint, graceful, and excellent fashion, yours is worth ten on't."
MUCH ADO ABOUT NOTHING, IV. 1.

CHARLES IX., who commenced his reign in 1560, was no sooner on the throne, than he assembled the States General at Orleans, to remedy the abuses introduced during the preceding reigns. Former sumptuary edicts were renewed, with added clauses. Women were prohibited from wearing gold ornaments on their heads, except during the first year of their marriage; the chains and bracelets they were permitted to wear, were not to be enamelled, under pain of a fine of 200 livres.

This, and numerous other stringent decrees on dress and ornaments, received no execution; luxury increased with the troubles and anxieties of the state. The disturbances which agitated the kingdom, calling the attention of the authorities to other matters, prevented the enforcement of the edicts. The fair sex lost no opportunity of showing their supreme contempt for right, reason, state policy, national requirements, or anything under the sun, that interfered with their caprice and vanity; and men were not behindhand in imitating their folly.

What has been said of the commencement of the reign of Charles IX., might be said with equal truth of that of his successor, who ascended the throne in May of the year 1574. The first two years of his reign were employed in the pacification of disturbances, and it was finally thought he had succeeded when he granted liberty of conscience and places of worship to the Protestants, by the Edict of Pacification, issued in May 1576. Among the abuses the King then undertook to reform, was the excess of luxury. The evil had taken enormous proportions; all conditions were confounded; the great consumption made of materials had greatly increased their price, while articles of food were proportionately high. The *ordonnances* of the preceding reigns were revived with penalties of 1,000 crowns (écus) to the infractor. It was moreover forbidden to all *roturiers* who had not been ennobled, to usurp the title or dress of a noble, and to their wives to wear the dress or the ornaments of *damoiselles*, or to wear velvet garments.

By an *ordonnance* of the 7th of September, 1577, this prince very expressly forbade all persons gilding or silvering on wood, leather, plaster, lead, brass, iron, or steel, saving for princes, or to adorn prayer books.

The same or similar *ordonnances* as those already mentioned were also issued, as usual, with the most minute directions as to how much or how little trimming, and whether on seams, hems, or elsewhere, each person was at liberty to wear. The tranquillity which the Edict of Pacification had restored, and which, during the course of eight years, was interrupted but once, by a war of a few months in 1580, gave ample leisure for the re-establishment of public order. In the different *ordonnances* issued, we always find particular attention paid to the specifying of the fashion of ladies' dresses and ornaments; for instance, the dames, and maiden damoiselles*, as also the wives of his councillors and their daughters, so long as they remained unmarried, were at liberty to wear pearls and gems set in enamelled gold, on their heads; ear-rings, necklaces, bodkins, rings, chains, bracelets, girdles, beads and chaplets, clasps and buttons in front of their gowns, and to the shoulder pieces (ailerons) and slits of their sleeves, limiting, however, these last ornamental fastenings to one row.

The number of rings, etc., etc., which women were entitled to, according to their rank, was also specified.

While the bourgeois were wisely restricted from incurring an expense in dress which they had not incomes to justify, the nobles indulged largely their taste for externals. The marriage of a Sieur de Vicour with a daughter of Claude Marcel, a favorite jeweller of Henry III., which took place at the Hotel de Guise, on the 8th of December, 1578, was honored by the presence of the court. On this occasion, the King made his appearance in an *entrée de ballet*, with thirty princesses and noble ladies habited in cloth-of-gold, and cloth-of-silver, and in white silk, enriched with quantities of gems of great value.

The cost of the court dresses was enormous, for pearls and precious stones were literally inwrought; gowns, doublets, mantles, coifs, glittered with gold and silver, and were laden with gems, excelling the marvels of fairy tales. The store of gold and silver plate, of magnificent jewels, was in accordance with the rank and fortune of the noble possessors—which, by the way, is not usually the case at the present day—and the head of the family had the use of them as he enjoyed the fief lands of his heritage; they passed as heir-looms to his descendants without alienation

* A damoiselle was a woman of noble blood, whether married or single.

or diminution. When a sovereign died, certain articles of value, which, according to traditional custom, became perquisites of the attendants, were ransomed by the heir. After the death of a high dignitary of the church, his plate and his *pomp* became the apanage of a church or convent, which always had a treasury for the reception of such articles. Jewels and plate were the last articles with which an illustrious house consented to part in days of distress.

The ladies who were distinguished by the favor of the French sovereigns during this century, contributed no little to the development which so many of the elegant arts attained in that favored age. Françoise de Foix, Countess of Chateaubriand, and Anne de Pisseleu, Duchess of Estampes, in the reign of Francis I.; Diana de Poitiers, Duchess of Valentinois, in that of Henry II.; Gabrielle d'Estrée. in that of Henry IV.; were all women of refined taste, excellent judges of art, and generous patronesses of talent. Diane de Poitiers especially was an amateur who made the most liberal use of her fortune. Her castle of Anet contained a magnificent collection of masterpieces in gold, silver and gems—perfect marvels of art. The munificence with which she rewarded artists of genius, was remembered long after every trace of her hospitals, and other beneficent institutions, had been swept away by time's wearing current.

Extravagance in jewels and dress was as great, if not greater, in the reign of the good Henry IV., as in that of his predecessors. The taste of the King would have led him to prefer simplicity in his own apparel, but his easy good temper, and foible for the fair sex, led him to yield to the prevailing mania, however opposed to his own inclinations.

Bassompierre tells us, that for the ceremony of the baptism of his children, the King had a dress made that cost 14,000 crowns; for the making alone he paid 600 crowns; the material was cloth-of-gold embroidered with pearls. He had purchased for the same occasion a sword, the hilt and scabbard of which were enriched with diamonds. The dress of the Queen Maria de Medicis, for the same occasion, was trimmed with thirty-two thousand pearls and three thousand diamonds. It was valued at sixty thousand crowns, but was so heavy, the Queen could not wear it.

At the baptism of the son of Madame de Sourdis, which took place on the 6th of November, 1594, Gabrielle d'Estrée appeared dressed in black satin, "so loaded with pearls and precious stones," says l'Estoile in his Journal, "that she could scarcely stand up."

The same writer adds shortly after: "On Saturday, November the 12th, I was shown a kerchief, a Parisian dealer in embroideries had just purchased for Madame de Liancourt (Gabrielle d'Estrée), and which said lady was to wear the following day at a ballet, and the price of it, as agreed upon, was nineteen hundred crowns to be paid on the spot."

Never had the Court of France been so magnificent as it was during the minority of Louis XIII. The number of lords attending it, and the peace the nation enjoyed, were motives for extraordinary extravagances; and the luxury displayed by the courtiers of the reign of Henry IV. was remembered with disdain by the beaux and belles of his son's court. It was in this reign that gold was first used on coaches, and to gild edifices.

The religious disturbances and civil wars which began in France in 1615, prevented the salutary check on luxury the different decrees issued would have otherwise enforced, and that which weakened the state, strengthened vanity and love of show. Edicts were ineffectual to prevent the continuance or the increase of the evil. Even the wealthy, straitened by ruinous extravagance, were induced to resort to disreputable means in order to keep it up. Imitation is contagious, and custom at last authorizes superfluities that in their origin had been considered the preposterous and absurd inventions of a few crazy individuals. When a fashion has gained ground and taken firm root, we find the wisest adopting that which they had decried. The firm authority, and especially the example, of the sovereign can alone check abuses and maintain the fashion of the day within the limits of good sense.

One of the greatest amateurs of gems of the close of the sixteenth and commencement of the seventeenth centuries, was the Emperor Rodolphe II. He had collected an immense quantity of precious stones, and had arranged them in a table so artistically that they represented a landscape as much like nature as a painting could have been.

The conquests of Mexico and Peru, achieved towards the close of 1543, and the dicovery of the rich mines there, made the precious metals much more common towards the commencement of the seventeenth century than they had hitherto been. Luxury, however, augmenting in proportion, this abundance, instead of decreasing the price of these metals, which were employed in immense quantities by jewellers and silversmiths, made it still higher; the gold mark was raised from one hundred and forty-seven livres to three hundred and twenty; and the silver mark from nineteen livres to twenty-five. The edicts of Louis XIII. at last put some restraint on extravagance in dress. Other wise decrees were found necessary to check the mad outlay in gold and silver plate.

Long experience has shown that of all laws, sumptuary laws are the soonest forgotten. No sooner was an edict published than the ingenuity of traders and manufacturers, encouraged by the vanity and love of ostentation of the nation, found means to elude its clauses, or invent some other extravagant whim unprovided against by the legislator. However ridiculous the empire of fashion, it is stronger than the wisest laws. The multiplicity of the sumptuary *ordonnances* issued during the long reign of the fourteenth Louis, far exceeded anything done during the preceding ones, and sufficiently indicates the progress of luxury, of manufactures, and of taste in the country where Fashion has established her seat.

Never had the passion for jewels and orfévrerie been carried to such a height as it was during the reigns of Louis XIII. and of Louis XIV. We are inclined to wonder, not only where all these gems came from, but where they all went to. France, once the richest nation in the world in jewels, is probably now one of the poorest in comparison with her former opulence. Successive revolutions, civil and foreign wars, have dispersed hither and thither the fine family-collections of former days. Of the nobles, who by voluntary exile escaped death, a very few carried their jewels with them; but none brought them back; they had been bartered for the necessaries of life. Men wear few jewels in the present age; in fact, they would ill accord with the sombre hue and ungraceful form of their garments; but with the satins, velvets, and brocades of the time of Louis XIII. and Louis XIV., they suited well. The accounts of the introducers of ambassadors can alone give an idea of what was spent by those sovereigns in gold-chains and medallions, in diamonds, *buffets* of plate, boxes, rings, and other articles, at every reception of a new ambassador. The diplomatic correspondence of those days shows that the most trifling negotiation cost France an enormous amount in presents of this kind, and her example was followed by every European power. But, though the coffers of the state were impoverished by this prodigality, the few individuals on whom these favors were lavished were not the richer for them, so great were the expenses their post entailed. In fact, the pomp and state displayed by each foreign envoy was inordinate, and usually absorbed all their resources. For the show made on the occasion of a first reception at court, we refer the reader to Lady Fanshaw's description of her husband going to that of Spain, as ambassador of England. In the reign of the fifteenth Louis, a Prince de Rohan not only spent his large income, but incurred debts to the amount of 600,000 livres, during his ambassy to Vienna.

The prodigality of princes and subjects gave a prodigious impetus to the trade of precious stones, especially to that of diamonds, which were preferred to all other gems. The innumerable court fêtes which were given during the reigns of Henry IV., Louis XIII., and Louis XIV., the ballets, comedies, masquerades, concerts, banquets, tournaments, and assemblies, given not only by the sovereign, but by the princes of the blood, the lords, and rich farmers-general, gave occasion, for the display of incredible luxury and magnificence. Cloth of gold and silver, borders of gems, costly lace, no longer sufficed to the extravagance of the courtiers; the material of which the dress was composed disappeared under the orfévrerie, enamel, and gems with which it was loaded. The women, in whose honor these sumptuous fêtes were given, tasked their invention to find new ornaments. The ferrets-pins with very rich and elaborately-wrought heads, and the aigrettes with which they adorned their hair were miracles of taste, skill, ingenuity, and richness. The inventory of the crown jewels, drawn up in 1618, describes these aigrettes, which may be also recognized in the designs of l'Empereur, the jeweller of the court of Louis XV. In addition to the diamond earrings, bracelets, necklaces, clasps, and aigrettes, diamond stomachers were introduced during this reign. Besides all these jewels, the Queen wore diamonds in her girdle, and diamond shoulder-knots. The actresses, in order to cope on the stage with the splendour displayed in the boxes, covered their regal robes with imitation gems that produced a very pretty effect.

Men were no less anxious to distinguish themselves in this respect than women. Knightly orders, sword and hat-knots, rings, shoe-buckles, waistcoat-buttons—all were profusely adorned with gems. All the diamonds and precious stones of a family were frequently sewn on a costume to be worn on some particular occasion. At a fête given by Louis XIV., in honor of Mademoiselle La Vallière, the King appeared as Roger in the ballet of *Alcides*, and wore a cuirass of silver, inwrought with gold and diamonds. He was mounted on a splendid horse, the caparison of which was scarlet, and perfectly dazzling with gold, silver, and precious stones. At this same fête, the Duke de Bourbon, who represented Roland, almost out-did the King. A quantity of diamonds were fastened on the magnificent embroidery that covered his cuirass and silk stockings; his helmet and horse-furniture glittered with diamonds.

From the above account, we may imagine what must have been the splendour of the King's costume when he appeared as the sun in the ballet of *Apollo*.

Among the fatal results to which the extravagant mania for jewels of that day led, may be quoted the case of Madame Tiquet, whose bridal-bouquet cost her her life as well as her fortune.

Carlier, a bookseller in the reign of Louis XIV., left at his death, to each of his children—one a girl of fifteen, the other a captain in the guards—a sum of 500,000 francs; then an enormous fortune.

Mademoiselle Carlier, young, handsome, and wealthy, had numerous suitors; one of these, a M. Tiquet, a councillor of the parliament, sent her on her fête-day a bouquet, in which the calices of the roses were of large diamonds. The magnificence of this gift gave so good an opinion of the wealth, taste, and liberality of the donor, that the lady gave him the preference over all his competitors. But sad was the disappointment that followed the bridal. The husband was rather poor than rich, and the bouquet that had cost 45,000 francs (£1,800) had been bought on credit, and was paid out of the bride's fortune.

The revelation of the deceit practised upon her, was not likely to ensure domestic peace; the lady, moreover, found that in lieu of living in the style she had expected, she would have to diminish her own expenditure to provide for her husband's. She soon solicited and obtained a separation and the use of her own fortune. The husband retaliated by bringing a charge of undue intimacy between his wife and Monsieur Mongeorge, a captain in the guards; and obtained from the king a *lettre-de-cachet* to confine her in a convent. Unfortunately for his plans, he could not forbear triumphing over his victim by exhibiting to her the fatal order; the lady sprang forward, snatched it from him, and threw it in the fire! Here was an end of his vengeance; forewarned is forearmed; the other side had probably partisans in power, and when he solicited a second lettre-de-cachet, it was refused.

During these little bickerings, the loving couple continued to reside under the same roof, but in separate apartments. This state of things was finally brought to a climax in a tragical manner; M. Tiquet one night received five stabs, of which, however, he did not choose to die—probably to spite his wife. The assassin was arrested, and confessed that he had been instigated to the deed by Madame Tiquet. The wife was beheaded; the servant, who had been the tool of her vengeance, was hung.

The gallants of the court of Louis XV. carried extravagance as far as the famous Egyptian queen. She melted a pearl—they pulverized diamonds, to prove their insane magnificence. A lady, having expressed a desire to have the portrait of her canary in a ring, the last Prince de Conti requested she would allow him to give it to her; she accepted, on condition that no precious gems should be set in it. When the ring was brought to her, however, a diamond covered the painting. The lady had the brilliant taken out of the setting, and sent it back to the giver. The prince, determined not to be gainsaid, caused the stone to be ground to dust, which he used to dry the ink of the letter he wrote to her on the subject.

England, meanwhile, was not behindhand in luxury of dress. The courtiers of Elizabeth, of James I., and Charles I., kept pace with their Gallic neighbours, and in the article of jewels worn on court dresses, especially by men, extravagance was such as no succeeding times have equalled. King James had a childish admiration for what was then called *bravery*, and to please their master's fancy, as well as to indulge their own vanity, his favorites exhausted their means, however ample. The attention paid to their personal decoration by the weak-minded monarch, is evidenced by the followin; extract from a curious letter in the British Museum, addressed to his son and to his favorite, then in Madrid, year 1623.

"I send you for your wearing the *three brethern*, that you knowe full well, but newelie sette, and the mirroure of France, the fellowe of the Portugal dyamont, quhiche I wolde wishe yo to weare alone in your hatte, with a little blacke feather.

"As for thee, my sweete Gosseppe, I send thee a faire dyamonde, quhiche I wolde once have given thee before, if thou wolde have taken it, and I have hung a faire peare pearle to it for wearing on thy hatte, or quhaire thow plaisis; and if my babie will spaire thee the two long dyamonts in forme of an anker, with the pendant dyamont, it were fit for an admirale to weare, and he hath enowgh better jewells for his mistresse—if my babie will not spaire the anker from his mistresse, he may well lende thee his rownde broachie to weare, and yett he shall have jewells to weare n his hatte for three grate dayes."

" Describing the different jewels, etc., which were sent to the Infanta, the king mentions "a head-dressing of two and twenty great peare pearls;" adding, "and ye shall give her three goodly pendant dyamonts, qwhair of the biggest, to be worne at a needle on the middeth of her forehead, and one in everie eare."

A court suit of King James' "Sweete Gosseppe," the Duke of Buckingham, cost £80,000.

The fashion of wearing jewels in the hat was universal throughout the courts of Europe at this time. The Spanish ambassador, Don Pedro de Zuniga, passing over Holborn bridge, his hat adorned with a rich jewel was stolen off his head, the mob openly encouraging the fellow who made off with it, because it was stolen from a Spaniard.

Lady Fanshaw, describing very minutely the dress of her husband going to be presented as ambassador at the Spanish court, says he had on a black beaver buttoned on the left side with a jewel of twelve hundred pounds value, a rich curious wrought gold chain made in the Indies, at which hung the King, his master's picture, richly set with diamonds . . . . on his fingers two rich rings, etc.

Sir Thomas More, in his *Utopia*, seems to ridicule the ornaments upon hats:—"When the Anatolian ambassadors arrived, the children seeing them with pearls in their hats, said to their mothers:—'See, mother! how they wear pearls and precious stones, as if they were children again.' 'Hush,' returned the mothers, 'these are not the ambassadors, but the ambassadors' fools.'"

CHAPTER VIII.

THE EIGHTEENTH CENTURY.

1775 to 1785.

The Diamond Necklace.—An unlucky Witticism.—Blood-royal in the Hospital.—A Confidential Friend of the Queen.—Royalty in Pecuniary Difficulties.—A Friend in Need.—The Invisible Purchaser.—A Prince's I. O. U.—The Queen's Billet-*due*.—Slight difference between Buying and Paying.—Disagreeable Morning Callers.—Starting for Church and going to Prison.

> " To her in haste: give her this jewel; say
> My love can give no space, bide no denay."
>
> TWELFTH NIGHT.

> " Win her with gifts, if she respect not words.
> Dumb jewels often in their silent kind,
> More than quick words, do move a woman's mind.'
>
> THE TWO GENTLEMEN OF VERONA. III. 1.

THE extravagance in dress and equipages which had continued during the successive reigns of four Louis, was put an end to by the Revolution of 1789. Louis XVI., who, in an age of extreme corruption, had remained pure in mind and plain in his habits, who had neither mistresses nor favorites, was unfortunately not of a temper sufficiently firm, severe, and imperious, to command fear or respect; and, even in his own household, was looked upon as a man of inferior abilities, a mere cipher, while his unaffected virtues exposed him rather to derision than admiration. The beautiful and gentle Queen had been brought up in the simple and economical habits which Maria Thérèse inculcated in her children, and which were then those of the German princes. The Empress herself, on occasions of public ceremony, displayed extraordinary pomp and magnificence, but in the routine of every-day life, she found time, even amid the cares of state, to play the good housewife. Unfortunately Marie Antoinette, transplanted when very young to a totally different atmosphere, and exposed to the fascinations of the most luxurious, as well as the most depraved, court of Europe, acquired habits of thoughtless expenditure, excusable in one so young and surrounded by such evil counsellors; but sadly misconstrued by the people, who, too far off to judge of causes, only saw the results. The wise lessons of economy practised in Vienna, were soon forgotten amid the more seductive splendours of Versailles. The suppers and the card assemblies of Marly presented the most brilliant spectacle, as full dress was required of all admitted to them.

In 1789, at a moment when the long-sown seeds of revolution were fermenting, when the political horizon was black with the impending storm, when famine was maddening the lower classes, the court, and especially the king's *maison militaire*, was displaying a magnificence so insulting to the needy, that it seemed as though a sense of the coming crash had turned their giddy heads. The officers loaded not only their uniforms with gold, but the harness and caparisons of their horses; the very manes and tails of the animals were plaited with gold braid. The singular infatuation that led the courtiers to brave an exasperated people, seemed to have infected the wisest heads, and the most generous hearts; and the mistakes made by Louis XVI. and his Queen at that dangerous crisis, had results as baneful as the excesses of their predecessors. Of these errors of judgment, one of the most important was, perhaps, the conduct held by these sovereigns in the affair of the *diamond necklace*—an affair that demands a special mention in the history of jewels.

This extraordinary case, which occupied the attention of the court of France, the Pope, the College of Cardinals, and the higher ranks of the clergy; the details of which were echoed throughout all Europe, casting a slur on the fair fame of a Queen, and finally proving one of the most deadly weapons in the hands of her enemies,—was neither more nor less than a swindle and forgery. Such cases are of daily occurrence, as the records of every criminal court will show. That which gave to this one its world-wide celebrity, making it one of the chief events of a century, the last twenty-five years of which were prolific in terrible catastrophes, was the rank of several of the parties concerned, the singular circumstances connected with the birth and life of others, the mystery thrown over some of the particulars of this astonishing fraud, the terrible passions called into play, the fearful results to the dupes and to the duper, and, above all, the extraordinary state of effervescence of the political parties of the day. The court, the parliaments, the clergy, the nobility, the provincial states, the people, all were in open hostility, and ready to seize on anything that could be used against their adversaries. The collision of these jarring elements gave to the unhappy affair of the necklace a terrible notoriety. Many versions of the facts have been given, and these have furnished ample materials to novelists; yet the following account, collated from all the documents of the case, from the memoirs, pamphlets, and petitions of the accused and the accusers, as they appeared at the time it was tried, may prove of interest.

In 1774, Louis XV., wishing to make a present to his favorite, Madame du Barry, commissioned Boehmer and Bassanges, jewellers to the court, to collect the most beautiful and perfect diamonds, and compose a necklace, that should be unique of its kind. Some time and a considerable outlay were required for such an undertaking. The French jewellers, resolving to furnish their royal customer with a masterpiece, made arrangements with their *confrères* at home and abroad to procure the largest, purest, and most brilliant stones. Unfortunately, before the necklace was completed, Louis XV. was laid in his grave, the throne was filled by a prince of a very different disposition, and the fallen favorite was fain to be content that she was permitted to retain the riches she possessed, without pretending to the execution of the deceased monarch's intentions. The work, however, was too far advanced to permit of its being abandoned without great loss, and hoping that Louis XVI. might be induced to purchase it for the Queen, the jewellers finished the necklace, which was valued at 1,800,000 francs (£72,000 sterling).

The low state of the new King's finances, the peculiar embarrassments under which he labored, being made responsible for the extravagance and folly of his predecessors, rendered such a purchase inopportune, if not impossible at that juncture. When it was offered, the young sovereign replied that a ship was more needed than a necklace; the costly ornament, therefore, remained on the hands of the manufacturers for some years, until the event which, breaking it up and dispersing the fragments brought together at such vast expense, gave it historical celebrity.

To understand by what a complication of circumstances a woman without position, fortune, favor at court, or even very great charms of person, could have conceived the idea of appropriating an ornament that was beyond the means of sovereigns, some knowledge of the position of the chief victim is required. The ascendancy acquired by a creature of astonishing boldness, cunning, and rapacity, over a weak-minded and credulous man, must be explained by reference to events much anterior to their meeting, and which gave rise to the life-long antipathy of Marie Antoinette, for Louis, Prince-Cardinal of Rohan.

In 1772, three years after the marriage of the Dauphin with the daughter of Austria, the Prince of Rohan was appointed ambassador to Vienna, and it was during his sojourn at that court, that he incurred the lasting displeasure of her who was destined to be his Queen. At one of the merry suppers of Louis XV., in the apartments of the favorite, the latter had drawn from her pocket, and read aloud, a letter purporting to be addressed *to her* by the ambassador at Vienna, and giving particulars of the private life of the Empress. Among other traits, characterizing the duplicity and covetousness of Marie Thérèse, she was described as drying with one hand the tears she wept for the dismemberment of Poland, while she extended the other to grasp her share of the spoils.

The insulting witticism that held up to the ridicule of the favorite and her guests the imperial mother of the Dauphiness, was, of course, immediately reported to the latter by one of the court butterflies; and the indignant daughter never forgave the imprudent writer.

Besides this over-true picture of Marie Thérèse, the Prince had been guilty of sundry other sins most heinous in the eyes of the prudish Empress; not the least of these was, the undisguised admiration he manifested for the fair sex. She vainly requested the recall of an ambassador, whose conduct set at nought her efforts to render her court a pattern of morality to all the courts of Christendom; Louis XV. viewed such offences in a different light, and the sinner was maintained in office.

The Prince was, however, guiltless of any thought of offending the Dauphiness; he had no correspondence, nor was he on terms of familiar intercourse with Madame du Barry; he had replied to enquiries of the king, who insisted on being made acquainted with all that took place at the Imperial Court. Louis XV. had left the ambassador's private letter in the hands of the Duke d'Aiguillon, and that minister, who was the creature of the favorite, had communicated it to her; she, with her customary thoughtless levity, had amused her guests with its contents.

Although Monsieur de Rohan, allied to the most powerful families of France, having relations in the most honorable posts about the King's person, and himself in the receipt of benefices constituting a princely income, had obtained, notwithstanding the enmity of the new Queen, the post of Grand Almoner of France, a Cardinal's hat, the rich abbey of St. Waast, and finally, had been elected Proviseur at the Sorbonne; the displeasure of the Queen, which entailed a sort of semi-disgrace at court, embittered his life. His enemies kept her anger alive by misrepresenting even the efforts he made to appease it. She had never granted him an opportunity to justify himself in person, and had obstinately refused to read any of the written vindications he had sent to her, although presented by the most influential of his friends. Though Louis XVI. was too mild and just to adopt extreme measures, he naturally shared the Queen's antipathy to a certain degree.

Such was the difficult position of the Cardinal at court, when a fatal chance introduced to his notice the intrigante, who, taking advantage of his well-known hobby—the desire to regain the royal favor—involved him in the disgraceful transaction that placed a Prince de Rohan, a Cardinal, a Prince-Bishop of Strasburg, Landgrave of Alsace, Prince of the Empire, Grand Almoner of France, Commander of the Order of the Holy Ghost, etc., etc., before the eyes of the world in the attitude of a thief and a forger.

The only part of this woman's statements ever proved to be true, was her claim of royal descent. Jeanne de St. Remy *de Valois*, was the daughter of the sixth descendant of Henry de Valois, second of the name, King of France, and of Nicole de Savigny, Lady of St. Remy. Jeanne's father, Jacques de St. Remy, appears to have been a thoughtless, weak-minded man. The domains of the barons of St. Remy, which, through the misfortune or extravagance of his forefathers, had dwindled down to a miserable remnant, disappeared entirely in his improvident custody. In 1760, the miserable man, reduced to beggary, burthened with a wife and three children, looking on his empty titles, unsupported by the lordly manors and broad acres from which they drew their names, as derisive humiliations, half crazed by pride, poverty, and shame, resolved to abandon his home near Bar-sur-Aube, and direct his steps to Paris. Having fastened his second daughter, three years old, in a basket, which he hung up at the window of a farmer, he set out, dragging after him his two other children, and his wife who was near her confinement. The unfortunate family received, on their arrival, such assistance as prevented their perishing in the streets: the mother gave birth to a fourth child, and the father, in a dying condition, was conveyed to the hospital of the Hotel Dieu, where the Chevalier Baron de St. Remy, of the blood-royal of Valois, ended his life.

Through the charity of the Marchioness of Boulainvilliers, the little orphans were clothed, fed, and educated. The boy was placed in the navy; the child born last having died, the one left in the basket was sent for, and taken care of by the same kind person who provided for the elder girl. In 1775, d'Hozier, having examined the family documents, testified to the genealogy of the young Valois; and the titles having been placed before the King, a pension of 800 livres was granted to the brother, then a lieutenant, and a like sum to each of the girls.

Jeanne de St. Remy married a gendarme of the name of Lamotte, whose father had risen from the ranks to the grade of lieutenant-colonel, and was killed at the battle of Minden. The valiant death of the father procured for the son a pension of fifteen hundred francs, which, however, did not prevent his contracting numerous debts. His marriage with the descendant of royal blood filled his mind with ambitious hopes, and prefacing his name with the title of Count, he came up to Paris with his wife, hoping to obtain new favors. Friends, however, began to weary of continual solicitations, and, after spending several years reduced to all sorts of shifts to keep up appearances that concealed a state of actual beggary, the wife presented herself before the Cardinal de Rohan, to petition that in, his capacity of Grand Almoner, he would procure some further aid for them from the royal bounty.

This dangerous siren, without being beautiful, had an intelligent and pleasing countenance: she was in the flower of her age, of middle height, with blue eyes, chestnut hair, and a fair complexion; she expressed herself with ease and elegance, and above all, with an apparent simplicity and frankness that carried conviction to the minds of her hearers. The Cardinal-Prince had a generous heart, more especially when the petitioner was a pretty woman; moved by the misfortunes of this descendant of kings, he advanced money himself and procured some few donations to relieve her immediate wants. The kind reception of His Eminence, and the ever-recurring wants of the applicant, encouraged the repetition of her visits. The lady soon fathomed the weak points of her protector. He advised her to apply in person to the Queen, and, lamenting it was not in his power to procure her an interview with royalty, he allowed himself to betray the deep chagrin the sovereign's displeasure caused him. This, he said, embittered every hour of his life. This confidential acknowledgment laid the ground for his ruin, the basis on which the fiend to whom he made it founded a scheme for which the annals of crime and folly have perhaps no parallel.

Some days after, Madame Lamotte returned with the news that she had obtained admittance to the Queen's presence, had been questioned kindly, had introduced the name of the Cardinal as being one of her benefactors, and finally, perceiving she was listened to with interest, had ventured to mention the grief he endured,—a grief which affected his health,—and had obtained permission to bring to Her Majesty his vindication.

That a man of such exalted rank and station, possessing the support of such powerful houses as those of Montmorency, Soubise, and Guémenée, could be brought to believe that an unknown, poverty-stricken woman had procured that which he and his friends had so long solicited in vain, seems the very height of credulity, but the folly of the Cardinal went far beyond even this: one who worshipped the charlatan Cagliostro as a god, was likely to fall into any snare—especially when his most darling wish was made the bait. The story of the tempter was plausibly told and well-fitted to work on such a man. "What thanks I owe you, my Prince, for the advice you gave me to present in person the narrative of my misfortune! I had gone to Madame* to solicit her kindly intervention; while there, the remembrance of all I had suffered overcame me to such a degree that I fainted. Her Majesty happened to come in at that moment, and, after manifesting the most generous sympathy, commanded that I should come and see her: a first visit brought about a second and a third; now I have an entrance into the private apartment, and I am led to believe I have the good fortune to interest Her Majesty, and that she deems me worthy of her confidence."

This *debût* had all the effect intended; the *protégée* became the protectress. Her reports were graduated with such art, that the Prince saw in her an intermediary who could not only pave the way for his return to court-favor, but obtain for him the highest post his ambition could aspire to. He instructed her how to manage to introduce his name often into the conversations she had with the Queen, and present to her, under the most favorable colors, the earnest desire he felt to be reinstated in the royal good graces.

Madame Lamotte's next report was extremely satisfactory. The Prince's apology, written by himself, containing all that could disarm the Queen's anger, had been placed in the sovereign's hands, and had met with so favorable a reception that a note was vouchsafed in reply—

* The King's aunt.

Madame Lamotte had previously ascertained that the Cardinal had not seen, or did not remember, the Queen's hand writing. The contents were as follows:—"I have seen your note; I am delighted to find you innocent. I cannot yet grant you the audience you solicit; as soon as circumstances permit, I will let you know. Be discreet."

These few words threw the dupe into extasies. From that moment he was completely blind. Convinced that Madame Lamotte was admitted daily into the private apartments, knowing her to be quick-witted and full of talent for intrigue, he thought it natural the Queen should be amused by her sallies, and that she would make use of her as a ready and officious tool. Following his guide's advice, he expressed his joy and gratitude in writing. The correspondence thus commenced was continued, and so worded on the Queen's part, that the Cardinal had reason to believe he had inspired unlimited confidence. When he was supposed to be sufficiently prepared, a note was risked containing a request which would have enlightened any other man as to the source whence it emanated. The Queen was desirous of assisting a family in distress; she had not funds by her at that moment, and commissioned the Grand Almoner to borrow 60,000 francs, and transmit them to her through the medium of Madame Lamotte. That a Queen, who had the Comptroller-General at her command, and could, especially for such a purpose, have had that sum from the royal treasury, should negotiate a clandestine loan, was something too absurd, yet the Cardinal never doubted. He borrowed the money himself and remitted it to Madame Lamotte, who brought, in return, a note of thanks.

A second loan of a like amount, for a similar purpose, was again requested, the term of payment being fixed in the note; and again did the Cardinal borrow of a Jew in order to oblige his condescending sovereign. The money certainly did go to relieve a distressed family; debts were paid and comforts procured; but the thief dared not allow her dupe to find too great a change in her Paris residence, although much that might have surprised him was accounted for by reference to the Queen's generosity. Wishing to appear in style among the inhabitants of Bar-sur-Aube, where she had so long resided in poverty, this artful creature insinuated to the Cardinal, through the usual medium,—a letter from the Queen,—that, in order to permit of Her Majesty's arranging matters so as to receive him publicly, without exciting suspicion, he would do well to absent himself for a short time. The Cardinal instantly set out for Alsace. He was no sooner gone than the two Lamottes, in elegant equipages, and attended by livery servants, departed on a visit to their old neighbours at Bar-sur-Aube. Here they furnished a house handsomely, displayed fine clothes, lace, jewels, and plate, and accounted, in a seemingly unreserved manner, for their opulence, by saying that the Queen's kindness supplied them with the means—Her Majesty would not allow a descendant of the Valois to remain in poverty.

Had the thefts gone no farther, they would, in all probability, soon have been discovered; but the victim, afraid of the ridicule his credulity would have excited, might, perhaps, have submitted to the pecuniary loss rather than make the affair public. But the soi-disant countess, emboldened by success, aimed now at much higher game.

The court-jewellers were by this time thoroughly tired of having so great a capital lying idle; they listened readily to the insinuations thrown out by an emissary of Madame Lamotte, that a lady having great influence at court might be able to recommend the purchase of the necklace. A visit to Madame Lamotte, and the promise of a rich present to the person who could negotiate this sale, was the result. The lady was cautious—she did not like to meddle with such matters—she would reflect on the subject. A few days after she called on the jewellers, and announced that a very great lord would go to them that morning, and look at the necklace, he being commissioned to purchase it. The Cardinal had, in the meanwhile, received from his royal correspondent, one of the little gilt-edged notes—contents as follows:—"The moment I am awaiting has not yet arrived; I hasten your return for a negotiation in which I am personally interested, and which I will confide to none but you: the Countess de la Motte will give you the explanation of this mystery." The signature of this note, like that of the former ones, ought to have raised doubts in the Cardinal's mind; the Queen always signed *Marie Antoinette;* the ignorant forger had added "*de France,*" which Her Majesty never did.

But the victim was blind. Back he rushed to Paris, delighted to prove obedience, astonishing his friends as much by the suddenness of his return, as he had by his unexpected journey, and little dreaming that he, a Rohan, was a mere shuttlecock in the hands of a creature who had lived on his alms.

No sooner was he arrived, than he was informed that the Queen earnestly desired the possession of the necklace; that wishing to purchase it without the King's knowledge, she would pay for it with money saved from her income. She had chosen the Grand Almoner to negotiate the purchase in her name, in order to give him an especial token of her favor and confidence. To this effect he was to receive an authorization written and signed by the Queen, which document he was to hold until he had been paid the money; that he was to make arrangements with the jewellers permitting of the payment being made in quarterly instalments, with the exception of the first ,which was not to be made until the 30th of July, 1785 (this was taking place in January). It was indispensable that the Queen's name should not appear at all in the transaction; the contract was to be made in the Cardinal-Prince's name only; the secret power signed by the Queen was to be a sufficient guarantee, and in all this he was to consider himself as receiving a signal proof of the confidence reposed in him.

The unsuspecting Cardinal hastened to fulfil his mission. In the middle of January 1785, the negotiation was commenced with the jewellers, and, after some discussicn, and the interchange of sundry other little gilt-edged notes, was concluded, the necklace being placed in the Cardinal's hands on the 1st of February. 200,000 livres of the original price were taken off, quarterly payments agreed to, and the Prince's note accepted for the whole amount. The jewellers were, however, made aware that the necklace was being purchased on Her Majesty's account, the Prince having shown them the power, and charged them to keep the affair secret to all *save to the Queen*. And most fortunate did it subsequently prove for the credulous negotiator that he did make this exception.

The day appointed for the delivery of the necklace to the supposed purchaser, was the eve of a great fête, Madame Lamotte having asserted that the Queen desired to wear it on that occasion. The casket containing the treasure was to be taken to Versailles, to the dwelling occupied there by Madame Lamotte, by whom it was to be remitted to the person the Queen was to send. The scene was well arranged and well played; the Cardinal, duly notified of the hour, arrived at dusk followed by his valet bearing the casket; he took it from the servant, at the door, and sending him away, entered alone. He was placed by Madame Lamotte in a closet the door of which opened into a dimly-lighted apartment, where the chief actress remained to play her part. A few minutes elapse—a door opens—a voice announces: "a messenger from the Queen"—a man enters—Madame Lamotte advances and respectfully places the casket in the hands of the last comer, who retires instantly. The trick is played, and the Cardinal is ready to swear that, through the glass sash of the closet door, he had perfectly recognized the confidential valet of the Queen!

Since the receipt of the 120,000 francs, Madame Lamotte had taken lodgings at Versailles, telling the Cardinal that the Queen, desirous of having her at hand, provided amply for her expenses there. In order to corroborate her statements, she would sometimes give him notice when the Queen desired her attendance at Trianon, and the infatuated man would disguise himself in order to see her when she went in, and join her when she came out. On one occasion, she was escorted some distance from Trianon by a man whom she reported to be the Queen's valet. The necklace was at that time neither purchased nor negotiated for; these were merely-stepping stones she was laying down. The soi-disant valet was a man of the name of Villette, a gendarme; he was the friend of Madame Lamotte, the comrade of her husband, and the useful accomplice of the pair, for he was the writer of the gilt-edged notes and of the power signed "*Marie Antoinette de France.*" To facilitate her scheme, Madame Lamotte had made the acquaintance of the concierge of the Château of Trianon, and it was his family that she visited when the Prince saw her go there.

This provident creature, who neglected nothing that could strengthen her victim's faith, had noticed that when the Queen passed from her own apartment, crossing the gallery to go to the chapel, she made a motion with her head, which she repeated when she passed the Œil de Boeuf. On the same evening that the necklace was remitted, she met the Cardinal on the castle terrace, and told him that the Queen was delighted. Her Majesty could not at that moment acknowledge the receipt of the necklace; but on the morrow, if he would be, as though by chance, in the Œil de Boeuf, Her Majesty would make such a motion of the head as would signify her approbation. But as to wearing the necklace, she thought best not to do so until she had an opportunity of mentioning its purchase to the King. A man predisposed to believe is easily convinced. The Cardinal saw the motion, and was satisfied it was intended for him.

The presence of the Cardinal proving troublesome, a little note again sent him to Alsace, and Madame Lamotte was able to give her attention to her affairs; the necklace was taken to London by her husband, and the sale of it, as it was, being too dangerous, it was taken to pieces; the smaller gems were reset in bracelets, rings, and other ornaments for the three accomplices, and the remainder sold to different jewellers. The money, having been safely invested, constituted a very comfortable rental.

In the meanwhile, the storm was gathering over the heads of the guilty, and of the merely imprudent. The Cardinal had urgently pressed on the jewellers the propriety of seizing the first opportunity that offered of seeing the Queen in private, to thank her for the honor she had done them; if they could not see her, they were to write. They did so, and were shortly after summoned to her presence to explain a letter that to her was an enigma. The whole affair of the purchase, at least as far as the Cardinal was concerned, was thus brought to the Queen's knowledge. This was in the beginning of July. From that moment. Marie Antoinette acted a part as unjust as it was undignified in a sovereign, bound to do justice to the meanest of her subjects without permitting her feelings to influence her decisions. So infamous a manœuvre, in which the Queen's name had been daringly used to obtain property to so large an amount under false pretensions, might well rouse her indignation, and should have prompted her to inform the jewellers that they were imposed upon, and had no one but the Cardinal to guarantee the payment. Had the Queen followed the dictates of reason and equity, she would have sent for the apparent culprit, and in the presence of the King and of the Prince of Soubise and Countess of Marsan, his relatives, have communicated the affair. Its real nature would have been disclosed, and the authors of the fraud punished. Had the Queen desired the humiliation of the Cardinal, she would have been satisfied, for he would have been obliged to resign his post at court, and retire to his diocese; this the sovereign had a right to exact; the powerful House of Rohan could have had no reason to complain; but there would have been no Bastille, no criminal suit, no *éclat*. Marie Antoinette, unfortunately, allowed herself to be guided by two of the most inveterate enemies of the Cardinal, and the consequence was an unexplained stain on her own name. The Baron de Breteuil, then minister, had long sought for an opportunity to crush the Cardinal; he convinced the Queen that His Eminence who, notwithstanding his princely income, was much embarrassed in his pecuniary affairs, had done all this in order to raise funds on the necklace. If immediately taxed with fraud, he might free himself from the charge by giving a flat contradiction to the jewellers' assertion, that he had acted in the Queen's name; this he could do, as they had no writing to the contrary, the contract being in his own name. If nothing was said, when the first payment became due, he was likely to compromise himself still farther, and might then be punished severely. Fearing to lose her full revenge, the Queen kept silence, leaving the Cardinal to the surveillance of his foes; and the jewellers were merely told to bring a copy of the agreement, and leave it with the Queen.

In the meanwhile, the greater part of the necklace had been disposed of, and the proceeds deposited in the Bank of England, in a fictitious name. In this the aid of an Irish capuchin friar, of the name of Macdermot, had been very useful. The date when the payment of the first instalment would be due was now fast approaching, and the Cardinal being wanted to provide funds for it, he was recalled to Paris in the month of June. A note assured him that he would soon see the realization of the Queen's promises; she was making great efforts to meet the first payment; the matter offered some difficulty on account of unforeseen expenditure, but it was hoped the funds would be ready in time. This was an artful hint that he was to prepare against any emergency. Three hundred thousand livres was, however, a large sum to be disbursed by a man who had already a debt to the Genoese of 600,000 livres, incurred when he was ambassador; another of 500,000 livres for the rebuilding of his palace; and one of 300,000 livres to the Jew, Cerf-Berr, 120,000 livres of which had passed into the hands of Madame Lamotte. The Prince began to think it very strange that no change whatever was apparent in the behaviour of the Queen in public; neither word, look, nor sign betokened a return of favor, and the necklace was never worn. It was evident that something must be done to satisfy him, and the inventive genius of Madame Lamotte got up a new farce. The Queen, for reasons left unexplained, not being able as yet to give him public marks of her esteem, would grant him a private interview, in which she would inform him of many things that could not be committed to writing. The meeting was appointed to take place between eleven and twelve o'clock on a certain night, in a grove at Versailles. The unexpected honor delighted the Cardinal; never was rendezvous awaited with more impatience.

Among the ladies of a certain class who frequented the promenade of the Palais-Royal, the Count de la Motte had met with one whose tall and elegant figure, gait, and even profile, gave her a great resemblance to the Queen. This girl, Mademoiselle Leguet, alias d'Oliva, was chosen by the Lamottes to personate the Queen in the scene they were plotting. The new accomplice was not initiated in the secrets of the worthy couple; she was merely told that the little part she was to act was got up to mystify a nobleman of the court; that it was for the amusement of the Queen, who would be an unseen witness of the manner in which she performed her part, and would cause her to be well paid.

Mademoiselle Leguet was brought to Versailles on the eventful day, and made to rehearse her part a few hours before it was to be acted, on the spot where the scene was to take place. She was instructed that a tall man, in a blue great-coat and slouched hat, would approach and kiss her hand with great respect. She was to say in a whisper—"I have but a moment to spare; I am greatly pleased with all you have done, and am about to raise you to the height of power." She was to give him a rose and a small box containing a miniature. Footsteps would then be heard approaching, on which she was to exclaim in the same low tone—"Here are Madame, and Madame d'Artois! we must separate!"

It was well known at Versailles that the Queen sometimes walked in the groves, with the ladies above mentioned, quite late in the evening, and the deception was grounded on this circumstance.

At the appointed hour, the Cardinal waited some time in vain for the Countess, who was to meet him on the terrace, and let him know if all went on well. His ally at last made her appearance, apparently in hot haste. She bade him not lose a moment; the Queen was much annoyed; she would not be able to prolong the interview, as she had intended, Madame and Madame d'Artois having proposed to accompany her: he was to hasten to the appointed spot, and Her Majesty would manage to escape from her companions for a moment and join him. The whole scene took place, as it had been planned; MM. de la Motte and Villette were the persons heard approaching, and who, being supposed to be the Queen's relatives, cut short the Cardinal's delightful tête-à-tête with royalty.

Fully convinced of his happiness, the blinded Prince complained bitterly to his confidants of the interruption that had prevented his opening his heart to the Queen; but he was more than ever resolved to prove his zeal. Understanding that she was suffering some anxiety on account of her inability to pay the 300,000 livres, he endeavoured to borrow them. A little note came to say it was impossible to furnish the whole of the sum; but that, if the delay of one month could be obtained, the jeweller should receive 700,000 livres at the end of August, in lieu of the 300,000 livres due in July: meantime 30,000 livres were tendered as amount of interest. This was all the Queen could do for the present.

Madame Lamotte made this sacrifice of 30,000 livres out of the large sums she had realized from the proceeds of her sales, in order to gain the farther delay which she needed to settle her affairs and retire from business. She began to feel very uneasy, and wished to get away; but she was not to escape so easily. The jewellers took the 30,000 livres, for which they gave the Cardinal a receipt, but they had been instructed to refuse all delay. They became very pressing; every morning they were at their debtor's bedside with the usual plea: they had made engagements counting on the payment of the first instalment to fulfil them, and have it they must. They finally spoke of making use of the power his note gave them. Urged beyond all patience, the Prince replied that he too would make use of his rights—they were about to make an *éclat* that would compromise the Queen, as he could only vindicate his honor by proving he was but her agent. "Why," he exclaimed, "since you have frequent access to Her Majesty, have you not mentioned the disagreeable situation in which her delay places you?" "Alas! Monseigneur," they replied, "we have had the honor of speaking to Her Majesty on the subject, and she denies having ever given you such a commission, or received the necklace. To whom, my Prince, can you have remitted it?" The Cardinal was thunderstruck; he replied, however, that he had himself placed the casket in Madame Lamotte's hands, and seen her deliver it into those of the Queen's valet. "At any rate," he exclaimed, "I have in my hands the Queen's authorization, and that will be my guarantee." "If that is all you count upon, my lord," said the jewellers, "we fear you have been cruelly deceived." From that day, they returned no more.

Madame Lamotte had gone to Bar-sur-Aube to secure her most valuable effects; being informed of what had been said by the jewellers, she hurried to Paris prepared, by a new lie, to vindicate herself, and, in case of the worst, to implicate the Cardinal still farther. She arrived at the Grand Almoner's in the middle of the night, her disordered locks and streaming eyes presaging some fearful disclosure. "I have just left the Queen," she exclaimed. "I have depicted to her the distress of Boehmer, the impossibility in which you are of satisfying him, and the scandal that would ensue. To my utter astonishment the Queen replied, that, since that was the case, she would deny having received the necklace, or authorized its purchase, and, to make good her own position, would have me arrested and ruin you. You see me crushed by this blow; I dare not return to my home, fearing it may be already surrounded by the Queen's emissaries; let me entreat your Eminence will give me shelter until I can concert with my husband the means of escape."

Again the Cardinal fell into the snare; it was improbable that she could have any share in the deceit, since she came and placed herself of her own accord in his hands. But in reality, this was a contrivance to clear herself, and criminate the Cardinal. She declared, when arrested, that the Prince had kept her a close prisoner for four-and-twenty hours, to prevent her disclosing her having been employed to sell diamonds for him; she had had great trouble to escape.

This new scene took place in the beginning of August. The state of mind of the Prince may be imagined. When he remembered the scene of the grove, the valet-de-chambre, the letters, he could not believe himself the dupe of the

Countess; yet the present conduct of the Queen, indicating as it did, a premeditated design to ruin him, seemed equally incredible. In the meanwhile, the Baron de Breteuil had drawn up a memorial to the King, in which the affair was depicted in the darkest colors. Appearances were bad enough, but they were blackened still more by the pen of an enemy. Care was taken not to present this memorial until the 14th, that the arrest of the culprit might take place on the 15th, which was a great fête day, when the Cardinal would officiate in the royal chapel in his pontifical robes, in the presence of the assembled court.

On the morning of the 15th, while the Grand Almoner was waiting to accompany the King to the chapel, he was summoned to the royal closet, where he found Louis and Marie Antoinette, the Baron de Breteuil, and two other court dignitaries. The King, handing him the depositions of the jewellers, and of St. James, a financier of whom the Prince had endeavoured to borrow for the Queen 300,000 livres, bade him read them: when he had finished, he asked him what he had to say to those accusations. "They are correct in the more material points, Sire," replied the Cardinal; "I purchased the necklace for the Queen." "Who commissioned you?" exclaimed she. "Your Majesty did so by a writing to that effect, signed, and which I have in my pocket-book in Paris." "That writing," exclaimed the Queen, "is a forgery!" The Cardinal, who still felt sure of what he asserted, threw a glance at the Queen, which probably conveyed more of his feelings than respect and etiquette permitted, and the King ordered him out. A few minutes after, he was arrested and sent to the Bastille.

A few days after, Madame Lamotte was arrested at Bar-sur-Aube. Among the singular features of this singular affair not the least is the fact of this woman remaining in her house, entertaining a numerous party of friends as tranquilly as though she had nothing to fear. She must have known of the arrest of the Cardinal on the 15th: she had taken care to send Villette and Mademoiselle Leguet out of the kingdom: her husband had made his escape on the 18th, and nothing had prevented her accompanying him. She may have thought that her accomplices being out of the way, she could brave the matter out, and fasten all the guilt on the Cardinal. This, however, as well as her cool insolence throughout the trial, and her exclamations against the Queen when her sentence was read, induced many persons to believe that she had been promised impunity, if she could throw all the odium on the Cardinal. This monstrous accusation, false as it no doubt was, the acrimony of the court against the Cardinal went far to strengthen.

Madame Lamotte was taken to the Bastille on the 20th. When examined, she at first denied all knowledge of the necklace, of its purchase or subsequent disposal, though she admitted that she and her husband had been employed by the Cardinal to dispose of a quantity of unmounted diamonds. She afterwards said, that the necklace had been purchased by the Cardinal, to sell in fragments, in order to retrieve his affairs; and that he had done everything with the connivance of Cagliostro, into whose hands the funds had passed. As for herself, she had never mentioned the Queen's name, or said she was admitted to her presence, etc. Though her statements were contradictory, her tone was ironical and daring beyond measure. In consequence of her accusation, Cagliostro and his wife were sent to the Bastille, where they were kept many months, but nothing proved that they had anything to do with the affair of the necklace. The Cardinal, who put the utmost faith in the cabalistic art of this great impostor, was wont to consult him, and probably the oracle had greatly corroborated the hopes he had founded on the Queen's favor.

The familiar intercourse that had subsisted between the Cardinal and Madame Lamotte was well known: how far the agents set to watch him, had initiated his enemies in the real facts of the case, has never been ascertained. The whole affair was, at this stage, apparently enveloped in mystery. The jewellers had treated with him alone; that the Queen had anything to do with the matter rested on his unsupported testimony, contradicted even by his accomplice, Madame Lamotte. The latter and Cagliostro charged each other spitefully, ripping up each other's lives in a manner highly amusing to the scandal-loving public, but neither explaining the mystery of the necklace, nor exonerating the Cardinal, whose position was very precarious, when light broke from an unexpected quarter. Father Loth, a neighbour of Madame Lamotte, whom she had imprudently admitted to her counsels, revealed to the friends of the Cardinal the part played by Villette and Mademoiselle Leguet in her affairs. Once on the traces of the intrigue, his partisans soon found both accomplices; Villette was arrested in Geneva, and Mademoiselle Leguet in Belgium. Their revelations changed the aspect of things. Investigations were made in England, and the sales of diamonds by Madame Lamotte proved. This, however, was not a point denied by Madame Lamotte; she represented herself and her husband as having been merely the agents of the Cardinal. But the evidence of Villette and that of Mademoiselle Leguet was conclusive as to the deception she had practised upon the Cardinal with regard to the Queen, and the other facts were easily proved.

The testimony of Cagliostro, notwithstanding her efforts to invalidate it by the disclosures she published of his real birth and antecedents, weighed heavily against her. Cagliostro being consulted on all occasions by Monseigneur de Rohan, had been made the confidant of his new hopes, and had been admitted to hear all that he supposed was doing in his favor from the lips of Madame Lamotte herself. The Prince's factotum, Baron de Planta, also swore to the same facts.

When confronted with these and other witnesses, and crushed by the mass of testimony against her, dropping her tone of daring, confident insolence, she gave way to ungovernable rage. "I see," she exclaimed, "there is a plot on foot to ruin me; but I will not perish without disclosing the names of great personages yet concealed behind the curtain!" This threat was not recorded on the register, but this wicked creature subsequently endeavoured to carry it out when her sentence was read to her.

During the course of the trial, an episode was brought to light which tended to complicate the proceedings, but which was finally put aside as foreign to the case. As illustrative of the cunning with which this female fiend had prepared her machinations, this intrigue is worth mentioning.

To clear herself in case of discovery, it was necessary to prove the Prince guilty, and to this end, some powerful motive had to be assigned for his conduct. Her fertile imagination was not at fault:—the Cardinal had a mistress, whose son needed a father; to marry the mother to a man who would lend his name to cover the past, a *dot* was needed,—his exhausted coffers could not furnish the sum; hence his fraudulent appropriation of the necklace! An adventurer, of the name of Bette d'Etienville was arrested at Lille, in Flanders, and brought to the Conciergerie in Paris. His revelations contained this extraordinary story: He pretended he had been employed by the Cardinal as a matrimonial agent to find a man of noble birth, who would marry a still young and beautiful woman, called Madame Mella de Courville de Selbark, the mother of a boy of fifteen. He had actually lured a certain Baron de Fages with this fable, and induced him to raise money and purchase jewels and fine clothes to a large amount, on the faith that he was to marry this demi-widow with the great fortune. The ramifications of this intrigue, in which the heroine was never seen by any one but the soi-disant agent, were numerous; the most romantic and improbable circumstances were related by this Bette. This counter-intrigue had been arranged and carried on by the provident Madame Lamotte, at the time she was deceiving the Cardinal; and the Baron de Fages was duped in the name of the charming widow *he was not to see until the wedding-day*, while the Cardinal-Prince was lured by the secret correspondence with the Queen.

This strange drama, that for more than a year had been acted before the Criminal Court, and before the world, was brought to an end on the 31st of May, 1786. As is usual in suits where the public appetite for scandal has been largely fed, even those who won had little reason to rejoice. The Prince-Cardinal was proved innocent of all fraud, but his extreme credulity had subjected him to the ridicule of the Parisians, and he was spared no mortification, either in prose or rhyme. Obliged by the King to resign his posts at court, he was unjustly exiled to his abbey of La Chaise-Dieu, in the mountains of Auvergne. The ungenerous conduct of the court, manifested not only in the manner of the Cardinal's arrest, but during the course of the trial, and even after his acquittal, gave rise to commentaries of the most injurious character.

The wretched woman who had raised this tempest was sentenced to be flogged, branded on both shoulders, and imprisoned for life. When the former part of the sentence was executed, she poured forth a torrent of foul abuse of the Queen; and, though she was immediately gagged, enough was heard and reported to form the ground of the vilest calumnies. Her husband, who had escaped to England, was condemned by default. He retaliated by the threat of the publication of a pamphlet, in which the Queen and the minister, Baron de Breteuil, were strangely compromised, if his wife were not set free. Such a threat, from such a quarter, would seem deserving only of contempt. Strange to say, however, Madame Lamotte was not only permitted to escape to England ten months after, but the Duchess of Polignac was sent across the channel to purchase the silence of this infamous pair with a large sum of money. It is very probable that the conduct of the Baron de Breteuil, the implacable enemy of the Prince-Cardinal, was not of a nature to be revealed to the world; but that the Queen, who could have nothing to fear from such a source, should have allowed herself to take a step so liable to misconstruction, proved very unfortunate. The bribe, too, was thrown away; for though one copy of the diatribe was burnt, a second was published some time after. A singular circumstance connected with this slanderous *Memoir of Madame Lamotte*, is, that the copies now extant in the Imperial Library of Paris were found in the Palace of Versailles, when it was taken possession of by the Republican Government.

CHAPTER IX.

THE EIGHTEENTH AND NINETEENTH CENTURIES.

The Revolution.—The Directoire.—The Consulate and Empire.—The Restoration.

The unfortunate affair of the necklace marked, under Louis XVI., the epoch of the decline of the inordinate passion for diamonds that had stamped the last two reigns. Those who possessed them wore them only at court. Dark clouds were gathering on the horizon. The unenviable distinction of a special costume for the Tiers-État, so injudiciously proposed by the advisers of the Crown, so indignantly resented by the class for whom it was intended, called forth the first attacks of the Assemblée, and was severely animadverted upon by Mirabeau, in his letter to his constituents. The Tiers-État, gaining the supremacy over the nobles and clergy, rushed into the opposite extreme. All distinctions of costume—the exterior tokens that had hitherto characterized the different classes of society—were abolished. The nobility, the clergy, the magistracy, and *the Finance*, lost all type. The levellers, carrying their furious zeal beyond all bounds, finally proclaimed their omnipotence, by compelling the monarch himself to submit to the disgrace of the odious red cap.

Amid these fierce antagonistic passions, taste and elegance vanished; the court parures of the ladies lost all distinctive character; the bastard style of the day was an affected mannerism, which was intended for uniform simplicity. A few rings and snuff-boxes, or bonbonnières, enriched with brilliants, and the singular appendages of two watches, one on either side, with each a huge fob-chain hanging down to the thighs, were about the only articles of jewellery worn by the beaux and the belles of the latter part of this reign.

Even this faint gleam of luxury was finally quenched in the revolutionary tempest that had well nigh annihilated every element of refined civilization, and restored the reign of barbarism, under the name of liberty.

With the Revolution, wealth and jewellery, competence and elegance, vanished entirely. Those who possessed these costly ornaments, carefully abstained from exhibiting articles that would have compromised the life of the wearer. Silver shoe-buckles were noted as tokens of aristocracy, and their place was supplied by ribbons. The few paltry trinkets worn were fashioned into shapes, and bore names to suit the bloody popular mania; ear-rings represented fasces, triangles, liberty-caps, guillotines; and were made of gold of ten or twelve carats. Even this base metal was too high for the assignats that were paid for it. Jewellers remained without work or resources.

The Reign of Terror ceased at last, and was succeeded by the Directoire. Shipwrecked society began to gather its waifs, and new agglomerations appeared at the surface. On one hand, was assembled the *jeunesse dorée*, consisting of the surviving shoots of ancient aristocratic families, or sons of the upper bourgeoisie; on the other, the class of *agioteurs* and army contractors, who found means to accumulate fortunes, amid the universal dearth of money, and who grew fat and insolent at the expense of the wretched soldiers, who were left by them without shoes or bread.

But even the diversions that were eagerly sought by all, still bore the stamp of the bloody ordeal through which the nation had so lately passed. Two of the places of amusement of the day—le Bal des Victimes, and le Concert Feydeau—acquired historical celebrity, as the scenes of the Renaissance of Luxury, and of a comparatively aristocratic society. To obtain admission to the Bal des Victimes, the candidate was bound to prove that he, or the patron under whose auspices he was introduced, had lost a relative by the knife of the guillotine. In addition to this lugubrious title, mourning garments, and hair *à la victime*, that is, cut close to the nape of the neck, as it was cut for convenience by the executioner, were *de rigueur*. But the mourning gradually became half-mourning, and this last was finally superseded by gay colors; and a few diamonds, in their ancient settings, were brought out of their hiding places.

The concert room of the rue Feydeau was more especially the rendezvous of the employés of government, of commissariates, of stock-jobbers. No one was excluded; and the exclusive class that constituted the Bal des Victimes mingled there with the newly-enriched men of the day. Luxury here took a new direction, and appeared under the most diversified origin and problematic education.

But it was under the influence and impulse of the Director Barras, that the renaissance of luxury took so bold a flight, and reached almost to the extreme of saturnalia. The republican government had given to manners, arts, and fashions, if not the classical impress of the Greeks and Romans, at least such a varnish as constituted a caricature of the ancients. The Directors, the Members of the Assembly, and the Five Hundred, were accoutred in Grecian caps and Roman mantles; their seats were shaped like curule chairs; even the *huissiers* were disguised as Roman lictors bearing fasces.

Women were not behindhand in the race backwards. Dresses, shawls, and shoes, had been superseded by tunics, mantles, and buskins. Diamonds and precious stones added brilliancy to the new costumes. These ornaments were scarce however, and were usually the spoils of ancient and noble families who sold them to repurchase the patrimony of their ancestors, or to obtain the re-entrance into France of some emigrated relative, or, perchance to procure the very means of existence.

But the setting of the few precious stones worn underwent a complete transformation. Parures of the reigns of the fifteenth and sixteenth Louis would scarce have harmonized with the costumes of the modern Aspasias and Julias. Jewellery and bijouterie took the Grecian and the Roman character. Diadems, clasps, rings, ear-rings, hair-pins, were all modelled on the antiques as seen in statues and graven stones.

The belles of the Directoire made the most of the few resources the scarcity of rich jewels left them. They attempted to compensate the deficiency of value by quantity and size. Those who could afford them wore antiques; those who could not, supplied their place with imitations. Cameos were much worn; they were set in necklaces, diadems, combs, pins, in clasps to fasten the sleeves on the shoulders, *à la grecque;* every ornament was set with an antique, or an imitation. It was at this epoch that the fashion of the rings on the toes was revived, and the classic fair ones promenaded the public gardens in Roman sandals that allowed them to exhibit their gemmed feet.

At the tables of the modern Luculluses of the day, another singularity was introduced, not however a revival from the ancients, but an invention of their own, quite in keeping with the mad fancies in which these republican autocrats indulged. At great dinners, it was the height of refinement that the salad should be mixed—not with the utensils commonly applied to such usage—but by the fair fingers of the loveliest dames present. Hence, it became indispensable that hands thus publicly exhibited should be embellished to the utmost, and the taper fingers that were plunged into floods of oil and vinegar, were studded with gems.

These absurdities might do for the leaders of fashion, but wealth had not yet had time to become disseminated, and the secondary classes were fain to content themselves with jewels of less value, and imitations. In the year VII. of the Republic, the material of the few trinkets that made their appearance was as poor as the workmanship; watch-chains, ear-rings, medallions, necklaces, brooches of gold enamelled black or blue, jet and coral ornaments, were set in base gold, and were all in the worst taste. Lapis-lazuli and cornelians were very fashionable.

In the year II. quantities of bracelets were worn, not as now, each one different, but in pairs, one on each wrist, and one above each elbow. They were not very expensive, however, as they consisted generally of several strands of coral beads. Pearls re-appeared under the Consulate.

Under the Empire, the art of the bijoutier began to revive slowly, though the prevailing mania continued to be the antique, or what was deemed antique. The splendid creations of the Assyrian and Etruscan art were not yet known; the rich Egyptian and Grecian ornaments were discovered at a later period. What was deemed classical simplicity was all the rage. *Armillæ*, in the form of serpents, plain rings, coral beads, scarabæi, and cameos, were deemed extremely chaste and beautiful. For nearly fifteen years there was a great demand for such articles. Pearls also re-appeared in *grandes parures*.

With the Restoration and the return of ancient families, the very few diamonds that had escaped the revolutionary storm re-appeared at court. These vestiges of the preceding century produced in the ladies a contrast similar to that which existed between the ancient *gentilshommes* of the court of Louis XVI., and who were designated under the name of *Voltigeurs*, and the men of the Empire, called by the uncourteous name of *Brigands de la Loire*. The fair one who proudly wore a *parure* Louis XVI. looked down on her contemporaries, adorned with Greek or Roman bijouterie, with a contempt only equalled by that with which a Holyrood general regarded the bearers of ci-devant imperial epaulets.

The novels of Sir Walter Scott not only caused a revolution in literature, but also in fashions. The passion for the old castles, dresses, furniture, and jewels described so minutely and vividly by the Scotch author, was prevalent among all classes. The Moyen Age put the Greek and Roman style to flight. In bijouterie nothing was seen but *chevalières, châtelaines, gibecières*, etc., etc. This fashion favored the introduction of colored gems. The revival of commercial intercourse, moreover, favored the importation of topazes, amethysts, beryls, and yellow crystals; all these low-priced stones were mounted in the most showy manner, and worn for *grandes parures*.

So poor in diamonds was France at that time, that the wedding present offered to the Neapolitan princess when she arrived as the bride of the Duke de Berry, was *strass*. When the trade of Paris presented to the Duke of Wellington a present of diamonds under £40,000, it was found necessary to borrow of the Civil List a certain number, to be returned subsequently in kind.

Under the Orleans dynasty, in order to set off to greater advantage the few diamonds extant, and make them compete with the large showy ornaments set with colored stones then worn, they were arranged in bouquets, with a great deal of silver around the gems, in order to increase the effect and make them appear larger.

At the present day, although the mediæval style has not wholly disappeared, it no longer predominates. In fact, the elegant art of the jeweller may now be said to be of the eclectic school; it borrows its designs from every country and every age, and the bijouterie of France is universally acknowledged to unite elegance, good taste, and variety in a higher degree than that of any other nation.

As a specimen of the style of jewels in 1828, our readers may be glad to find here an inventory of those of Mademoiselle Mars, one of the richest collections possessed by a private person at that time. This catalogue was published at the time of the robbery of the diamonds of that celebrated actress, with a view of discovering the thief, and so exact was the description, that every stone was identified, though all the settings had been melted. Should a copy of this work, escaping the destructive touch of time be preserved to future ages, the antiquaries of posterity may be grateful for the preservation of one of those precise, positive, and technical documents, which we have too often reason to regret our ancestors have neglected to leave to us.

1 Two rows of brilliants set *en chatons*, one row composed of forty-six brilliants, the other of forty-four; eight sprigs of wheat in brilliants, composed of about five hundred brilliants, weighing fifty-seven carats; a garland of brilliants that may be taken to pieces and worn as three distinct ornaments, three large brilliants forming the centre of the principal flowers, the whole comprising 709 brilliants, weighing eighty-five carats three-quarters; a Sévigné mounted in colored gold, in the centre of which is a burnt topaz surrounded by diamonds weighing about three grains each, the drops consisting of three opals similarly surrounded by diamonds; one of the three opals is of very large size, in shape oblong, with rounded corners; the whole set in gold studded with rubies and pearls.

2 A parure of opals, consisting of a necklace and

Sévigné, two bracelets, ear-rings, the studs of which are emeralds, comb, belt-plate set with an opal in the shape of a triangle; the whole mounted in wrought gold, studded with small emeralds.

3 A gothic bracelet of enamelled gold, in the centre a burned topaz surrounded by three large brilliants; in each link composing the bracelet is a square emerald: at each extremity of the topaz forming the centre ornament are two balls of burnished gold, and two of wrought gold.

4 A pair of girandole ear-rings of brilliants, each consisting of a large stud brilliant and of three pear-shaped brilliants, united by four small ones; another pair of ear-rings composed of fourteen small brilliants forming a cluster of grapes, each stud of a single brilliant.

5 A diamond cross, composed of eleven brilliants, the ring being also of brilliants.

6 A bracelet with a gold chain, the centre-piece of which is a fine opal surrounded with brilliants; the opal is oblong and mounted in the gothic style; the clasp is an opal.

7 A necklace of imitation pearls, the clasp set with a large brilliant; a couple of pear-shaped emeralds, the studs set in rose-diamonds.

8 A necklace of pale-colored emeralds, surrounded by precious stones: the stones are not all well shaped; at each end of the necklace a few false emeralds have been added in order to lengthen it; a pair of ear-rings, assorted girandoles.

9 A parure of small rubies, several of which have dropped from their setting; to the necklace a small cross has been added; a comb, the rubies of which are *moulé à jour;* the ear-rings consisting of four small rubies, the whole very simply mounted.

10 A bracelet, with five graven *nicolos* mounted on square gold plates, with small ornaments in the four corners and wrought gold links.

11 A bracelet *bonne-foi*, with a serpent-chain forming serpents, and a ring fastened by a chain to the bracelet.

12 A gold bracelet, with a *grecque*, surrounded by six angel heads graven on turkoises, and a head of Augustus.

13 A dog-collar bracelet, the clasp a white cameo on a sardonyx ground, representing a female head.

14 A belt plate, a cameo on a brown ground, representing the head of a female mounted in gold.

15 A serpent bracelet *à la Cléopâtre*, enamelled black, with a turkois on its head.

16 A bracelet with wrought links, burnished on a dead ground, the clasp a heart of burnished gold with a turkois in the centre, graven with Hebrew characters.

17 A bracelet, with a row of Mexican chain and a gold ring set with a turkois and fastened to the bracelet by a Venetian chain.

18 A small purse of wrought gold, garnished with small rubies, also gold tassels; the purse opens by means of a chain *de jaseron*.

19 A large oval emerald pin, the mounting plain, the emerald deep-colored but with flaws.

20 A gold chain, oval links, enamelled with small dead gold links.

21 A Brazilian or Mexican chain.

22 A ring, the hoop encircled with small diamonds.

23 A ring, the hoop set round with pearls.

24 A ring, à la chevalière, set with a square emerald between two pearls.

25 A gold chevalière ring, on which is engraved a small head of Napoleon.

26 A small chevalière ring, set with a cabochon turkois.

27 A small wrought-gold chevalière, the stone—a small oval hyacinth—is lacking.

28 A serpent ring, *à la Cléopâtre*, of burnished gold, with a small square emerald in the centre.

29 A small emerald and pearl ring.

30 A bracelet, composed of four gold plates à charnières.

31 Two Grecian bracelets of imitation gold.

32 A pair of ear-rings, gothic style, colored enamels.

33 Two large studs of false rubies, mounted in imitation gold.

34 A cross, *à la Jeannette*, with heart and ear-rings to correspond.

35 Two belt buckles, gothic style, one of burnished gold, the other set with emeralds, opals, and pearls.

36 A cross of dead gold set with colored enamels.

37 A little ring with a false turkois, engraved with a head of Napoleon.

38 A small flacon of wrought-gold, form—a bonbonnière.

39 A square cassolette, rather large, garnished with small rubies.

40 A necklace of two rows coral; a small bracelet of engraved cornelians.

41 A comb of rose diamonds, form D 5, surmounted by a large rose surrounded by smaller ones, and a cinque-foil in roses, the chatons alternated, below a band of roses.

42 A necklace of nine *plaques* de brilliants, the centre one the largest; said plaques united by a chain of roses.

43 A pair of brilliant hoop ear-rings, with a brilliant surrounded by smaller ones in each stud.

44 A necklace of imitation pearls, the clasp two large brilliants united by a small one.

PART SECOND.

THE GEOGRAPHY OF PRECIOUS STONES.

CHAPTER I.

ANCIENT FICTIONS AND MODERN DISCOVERIES.

The Mythological Origin of many Gems.—The Diamond, the Amethyst, the Pearl, the Emerald, Lapis-Lazuli, Amber.—Ignorance, or at least silence, of the Ancients with regard to Mines of Precious Stones.—Jealousy of Ancient Traders in Gems.—The Griffin-guarded Emeralds of old.—Egyptian Mines of the Ptolemies brought to light in the Nineteenth Century.—Demon-guarded Mines of the Present Day.—Emerald Mines in the Tyrolese Alps.

In the preceding chapters gems have been spoken of as already wrought and set by the artist to supply the requirements of luxury, and adding by their lustre to the magnificence of the temples, altars, and statues of the Deity, adorning the crown of monarchs, and giving éclât to rank and beauty. The origin of these beautiful productions, and under what conditions of soil, climate, and labor, nature yields them to the eager search of man, remains to be told.

The origin of diamonds and precious stones, that is, the first discovery of them by man, is lost in the night of ages. History affords no data on the subject. Among the ancients no traveller, no naturalist, no historian, has left us a description of mines of precious stones. Many ancient writers speak of the gems, and mention the purposes to which they were applied; but none have troubled themselves to furnish preliminary notions with regard to their geography, or birth-place. The imagination of poets, never at a loss to account for everything in nature, as well as for every invention of human ingenuity, undertook to supply that which the science of philosophers had omitted to give; they provided every valuable gem with an illustrious, as well as marvellous, origin. Thus, Diamond was the name of a beautiful youth of the island of Creta, one of the attendants of the infant Jupiter in his cradle. The god who placed among the celestial bodies the nymphs and goat that had reared him, could not leave this youth subject to "the ills that flesh is heir to;" Diamond was transformed into the hardest and most brilliant substance in nature.

Aristotle adds the weight of his authority to the origin poets have given to the amethyst. A beautiful nymph, beloved by Bacchus, invoked the aid of Diana, who answered the appeal by metamorphosing her votaress into a precious gem. The baffled god, in remembrance of his love, gave to the stone the color of the purple wine of which he had taught mortals the taste, and the faculty of preserving the wearer from its inebriating effects.

The ignorance that prevailed during so many ages, as to the real nature of amber, gave rise to innumerable fictions. Nicias, the historian, asserts that the heat of the sun is so intense in some regions that it causes the earth to perspire, and the drops coagulating, form the substance called amber. These drops of perspiration were carried by the sea into Germany. Pliny asserted it to be a superabundance of the sap of certain pine trees. Poets give several versions of the origin of amber. According to some, the Heliades, sisters of Phaëton, though metamorphosed into poplars on the banks of the Po, still lamented the death of their brother, and wept tears of amber. This, however, is very gravely contradicted by Theophrastus, who asserts, that Phtaëon perished in Ethiopia, near the temple of Jupiter Ammon. Sophocles says, that amber-drops are the tears of the Meleagrides, the sisters of Meleager, changed into birds, and weeping their brother's loss. The Gauls again accounted for amber as being the divine drops that fell from the eyes of Apollo when, grieved at the death of his son Æsculapius, and that of the nymph Coronis, he forsook Olympus to dwell among the pious Hyperboreans. Eastern poets say that it is a gum from the tears of certain consecrated sea-birds. To this fanciful origin a celebrated English confrère of the lyre has alluded :—

> "Around thee shall glisten the loveliest *amber*
> That ever the sorrowing sea-bird hath wept;
> And many a shell in whose hollow-wreathed chamber
> We, Peris of Ocean, by moonlight have slept."

A fanciful abbé was of opinion, that amber was honey melted by the sun, dropped into the sea from the mountains of Ajan, and congealed by water.

The ancients were greatly puzzled to account for the presence of the pearl in the oyster ; but they finally arrived at a satisfactory conclusion. The oyster, at certain seasons, opened to receive the dew—the pearl was the offspring of this union. The pearl was large or small, and more or less pure and beautiful, according to the size and purity of the dew-drop the oyster had received in its bosom.

Lapis-lazuli has also its legendary origin in India. A curious native treatise on various subjects of natural history, called Calpayueti, has the following :—

"From a cry of the giant son of Diti, resembling the roaring of the troubled ocean at the close of the Calpa, sprung the variegated Vaiduryam (lapis-lazuli) source of colors, of a bright and ravishing splendour. Not far from the declivity of Mount Vidura was the mine of that precious stone, but limited to particular seasons for its production, and then closed. First, from the origination of that demon cry, did this mine suddenly spring into the world, eminent in its properties, the ornament of the three worlds; but ever since, on the muttering of the clouds of the rainy months (July and August), imitating the sound of that prince of demons, are those beautiful vaidurya gems emitted, of varied lustre and rapid effulgence, as of a multitude of fiery sparks*."

The situation of Vidura is identified with that of Mount Rohana, or Adam's Peak, in Ceylon.

There is considerable difficulty, if not impossibility, in identifying the gems mentioned by the ancients. In the Scriptures, all the precious stones known to us are supposed to be mentioned. The vestments of the High Priest glittered with jewels; and, it has been conjectured that the two onyx stones, "enclosed in ouches of gold, graven as signets, are graven with the names of the children of Israel," that were placed on the shoulders of the ephod, were in reality diamonds. Admitting this to have been the case, we also admit that the art of cutting the diamond, supposed to be of modern invention, was known to the Hebrews of that day as well as it now is to their Dutch descendants. Whether the stone called shamir, which the Sept. in Jer. xvii. 1, and the Vulgate, in all the passages where it is mentioned, take for granted was the diamond, was really the stone to which that name is now given, is a question yet unsolved.

We labor under the same uncertainty with regard to other gems. The ancients gave the name of *smaragdus*, which we take to have been the emerald, to stones of a very different kind. Theophrastus speaks of an emerald four

* Oriental accounts of the precious minerals, translated by Rajah Kalikischen, with remarks by James Prinsep, F.R.S., in the Journal of the Asiatic Society of Bengal. Calcutta, 1832.

cubits high and three broad, sent by the King of Babylon to the King of Egypt, Ptolemy Philopator, for the statue of his wife Arsinoe; he also mentions an obelisk forty cubits high, composed of forty emeralds. It seems probable that Theophrastus speaks of green jasper, or of emerald prisms, or again of the green crystallizations found at the mouth of volcanos, or in the lodes of mines, and which, though heavy, are soft.

The description given by Pliny of the emerald only suits the peridot, a yellowish green stone, found in the island of Cyprus, and which has also been denominated the bastard emerald.

The sapphire of the ancients is supposed by mineralogists to have been nothing more than our lapis-lazuli.

We will leave to our scientific writers the task of reconciling ancient names with modern ideas. The silence maintained by ancient authors on the birth-place of gems, may, in part, be accounted for, by the extreme jealousy with which the nations who traded in them, endeavoured to conceal the sources whence they drew this rich merchandize. When the truth could not be wholly concealed, it was disfigured by the most absurd fables, in order to deter adventurers from establishing a competition, and also, perhaps, to add supernatural value to the real one of the gems. The Syrian, or finest oriental emerald, was reported to lie in gold mines, the access to which was attended with great risk and difficulties, as they were guarded by ferocious and wicked griffins, who, led by a singular instinct, were constantly at work in the bowels of the earth, searching for gold and precious stones, which, when they had found, they never gave up but with their lives. Nor could the gems be found, or forcibly conquered, from their strange keepers, by ordinary mortals; they could only be procured by application to the Arimaspes, a nation of pigmy Cyclops, who sallied in armed bands to fight with the griffins, of whom they were the born enemies, and to despoil them of the treasures they were constantly gathering. The Arimaspes were by some supposed to inhabit Scythia, by others, the Riphean mountains. Such fictions as these being sanctioned by the testimony of serious writers, such as Pliny, Pomponius Mela, Strabo, and Pausanias, it cannot be wondered at, that the darkness which enveloped the real facts, should so long have remained unbroken.

Strabo asserts that death awaited the stranger who navigated towards the island of Sardinia, or the Pillars of Hercules,—he was sure to be drowned. This report was circulated by the Carthaginians, who, as may be seen by their treaties with Rome, were extremely jealous of any approach to an island whence they obtained the sardonyx, with them an important article of trade. By the Pillars of Hercules, must be understood the south-west of Spain, where they possessed such rich gold mines.

According to Heeren, the Etruscans and Carthaginians carried on a large trade in diamonds and precious stones, which they obtained in part from the interior of Africa.

The fact, that neither among the graven stones of the Greeks and Romans, among the presents made to churches, nor the old treasures of kings, had a single real emerald been found until lately, induced mineralogists to conclude that this gem was not known in Europe, previous to the discovery of America. Others, however, do not think that because the emerald mines of the Greeks and Romans are lost to us, we are justified in denying their existence. We are still little acquainted with the mineralogical wealth of Asia, Africa, and the Archipelago. We have not yet found the beds of those large sardonyxes, on which the Romans engraved such fine cameos, neither have we yet discovered the copper mines, once so numerous in the same regions; we should, therefore, hesitate before passing too severe a judgment on the mineralogical knowledge of the ancients.

Setting aside the engraved emeralds mentioned in history, and on which some discussion might be raised, the existence of emeralds in the treasuries of ancient basilicas, long before the discovery of the New World, cannot be denied, since that which adorns the tiara of the present sovereign Pontiff, and which was presented to his predecessor, on the occasion of his visit to Paris, in 1804, bears the name of Pope Julius II., who died thirty-two years before the conquest of Peru. By what means this stone found its way from the treasury of Rome to the empire of France, we will not attempt to explain; but its existence would alone suffice to decide at once the question of the emerald having existed in Europe before the conquest of Peru. All doubt, however, is now at an end, by the recent discovery of the famous emerald mines of Mount Zebarah, hitherto only known to modern times by the mention made of them by ancient writers, and the confused report of the Arabs. Monsieur Caillaud, the persevering mineralogist to whom the scientific world is indebted for the solution of this long-discussed question, found these mines nearly in the same state in which they had been left by the engineers of the Ptolemies: a multitude of excavations, and subterranean canals, carried to a great depth, and admitting of four hundred men being employed therein, extensive causeways, and other important works, show on what an extensive scale these mines had been worked. The ropes, baskets, levers, grinding stones, vases, lamps, and other tools and utensils, were still lying around, as though the miners, over whose ashes several centuries had passed, had left them there but yesterday. The mode of working of the ancients is now ascertained.

This valuable discovery confirms in all points the testimony of Strabo. Having described the isthmus, that is the space, narrower there than elsewhere, that separates the Nile from the Red Sea, he adds, "On this isthmus are situated the mines of emeralds, and other precious stones, which the Arabs extract by means of very deep subterranean canals." This most precise and clear description is completely verified by the recent visit of Monsieur Caillaud.

When Monsieur Caillaud first found these mines, he had with him but seven men. On his return to Cairo, he presented to the Viceroy a fine specimen of the riches that existed in his own domains, ignored by their owner; and he was immediately requested to return there and re-open the mines. This second expedition, undertaken under the special protection of the Pasha, who furnished him with a numerous company of armed men and miners, together with camels, and every necessary, lasted two months and a half. The following extract from M. Caillaud's report, describing his first descent alone into cavernous depths unexplored by the foot of man for several ages, is another proof, added to the many we have, that the love of science is as powerful as that of glory, to make men brave every peril, endure every fatigue and privation. His guides had told him there were immense subterranean caves in the mountains.

"Having reached these caves, I knew them at once to be mines; what kind of mines I could not yet tell; I saw only lodes of mica, talc, and schistus, interrupted by the masses of granite forming the body of this mountain. I set three A'babdehs to clear the entrance to one of the excavations. As I was resting from the fatigues undergone during that day and the preceding ones, my eyes chanced to light on a fragment of emerald of a dark green color. My surprise and delight were great. Forgetting all fatigue in my impatience to visit the level, I encouraged the A'babdehs, and began to work with them; we soon gained an entrance into the mine. I immediately caused torches to be lighted, and, accompanied by my interpreter and an A'babdeh, I descended by a very oblique road. I had scarcely gone a hundred paces, when I found the too rapid inclination of the lode rendered the road dangerous. The frightened A'babdeh turned back; my interpreter, finding the way too narrow, hesitated and stopped short; I alone continued to descend for the space of three quarters of an hour; I then found the road choked and blocked up by enormous masses of mica, which had fallen in from the top; I was alone to clear the way; I had gone a distance of four hundred feet under ground, through many difficult and even dangerous passages; my strength would not permit my undertaking to remove the obstacles, and I was obliged to give up my attempt to proceed farther. I was about to re-ascend, disappointed at having made no discovery, when amid the masses of mica I perceived an hexahedral emerald prism; I detached it carefully, leaving it in its gangue. I wandered some two hours longer in these narrow levels, during which time my interpreter began to feel alarmed for my safety, the great depth I had reached quite preventing my hearing his reiterated calls; he sent for a rope and let it down into the shaft, thinking it might reach and be of some assistance to me to return, but none of my men dared venture down. My light beginning to burn dimly, after resting a little, I again sought the upper road, the ascent of which I found very laborious. Amid the profound silence that reigned around me, the voice of my interpreter at last reached my ears, and guided me to the spot where he stood. His first question was, had I found many emeralds? I replied in the negative, but in such a tone that he was fully persuaded I had my pockets full; this thought was a greater punishment to him than any reproaches I could have addressed to him."

The following day, our persevering mineralogist found more than forty excavations, such as the one described. These mines, abandoned many ages, are probably for the majority, filled in by the falling of portions of the mountain, and by the stones brought by the torrents. The mountain in the vicinity of this one (Mount Zebarah) is also filled with excavations which extend to a great depth. The Egyptian emerald is embedded in lodes of black micaceous clay slate, which penetrate the masses of granite of which these mountains are formed. It is also met with in the accidental cavities of several granites; the finest and purest are found in hyaline quartz.

It is a fact that the Arabs entertain to this day the same superstitious fears with regard to these mines that the ancients did, with the slight difference that mythological belief established. A deputation of the Arabs of that district, who came to learn the reason of M. Caillaud's visit there, cautioned him strongly against sleeping near the caves, as they were very dangerous, being the refuge of snakes, wolves, and other beasts of prey; and the abode of demons who would resent the intrusion.

So fully persuaded were the men of the suite of M. Caillaud of the truth of this account, that none of them closed their eyes that night, and they spent their time firing off their guns to keep off the evil spirits of which the A'babdehs had spoken.

It was in his second expedition, in the following year, that M. Caillaud ascertained that some of the excavations had been carried to a depth of 800 feet below the level; some of them were so spacious that 400 men could work there at once. Seven leagues from Mount Zebarah, he found mountains containing much larger emerald quarries; and some, in which there were more than a thousand excavations, furnished facilities for communication between the upper and the lower ground workmen, on an extensive scale; there were stone causeways, along which the camels could convey provisions to the farthest extremity of the mines.

About half a league from these quarries, M. Caillaud discovered a Greek tower, and five hundred houses, still standing in very good condition, though probably they had not been inhabited for two thousand years.

"The ancients appear to have cared little about facilitating the labor of their mines, as these mines would now be looked upon as impenetrable. The miners were obliged to creep or slide, sometimes on hands and feet, sometimes on their backs, or quite flat on their chests, through little narrow paths running in all sorts of oblique directions according to that which the lode of talc, mica, or clay slate followed, sometimes going in this laborious way a distance of four or five hundred feet. Where the schistus happened to exist in large masses, excavations were made in which several hundred men could work: thence again a hundred new paths would be run off, extending to vast depths, and forming inextricable labyrinths."

The explorers were compelled to give up the undertaking for that year, having been disappointed in the periodical rains expected at that season. After searching in vain for the wells that must have been known to the former miners, in Mount Zebarah, M. Caillaud was compelled to retrace his steps. From one hundred and fifty men, and a corresponding number of camels, he had been reduced to thirty, having sent the remainder to places where they could procure water.

"What," he exclaimed, "could I expect to accomplish in these immense quarries, with thirty men, where five thousand might be employed in clearing."

According to tradition, Aly Bey worked a portion of these quarries about a century ago. M. Caillaud easily recognized four excavations that had been worked as recently as that, the action of the air having given a much deeper tint to the talc and schistus which the ancient miners had extracted; there were also ruins of dwellings in the valley of Zebarah, the fragment of a mosque, Arab inscriptions, and Mussulman tombs, belonging to a late period.

M. Caillaud brought back to the Pasha some ten pounds weight of emeralds taken from the mines of Mount

Zebarah. These stones were, with some exceptions, of a pale green, cloudy, and full of flaws. This kind of emerald is well known to the trade, in Cairo or Constantinople, either in large pieces, or in small ones, perforated for ear-rings. The harness of the Sultan's horses is also covered with similar emeralds taken from Egyptian quarries.

The emerald may be found in all countries where the soil is granitic. At Adoutschelon, in Siberia, they are embedded in hyaline quartz, forming veins in the graphite granite.

The first emeralds known proceeded from the mountains of Africa, situated between Ethiopia and Egypt. The emerald of Pope Julius II. probably came from thence. This emerald is in the shape of a short cylinder rounded at one of its extremities; it measures about twenty-seven millimetres in the direction of its axis by thirty-four in diameter. The African emeralds are far less precious than the Peruvian; their tints are less pure, and they frequently contain foreign substances which render their reflection changeable. Thus, notwithstanding the discovery of the mines of Ptolemy, the American emeralds retain their high value, and it is supposed that the African mines whence the ancients drew their splendid stones are still unknown to the moderns. Pliny mentions Scythia as the country whence, in his days, the best were brought.

The Tyrolean Alps, in the neighbourhood of Ried, (kingdom of Bavaria,) are rich in emeralds of rare beauty; we have seen specimens of the finest dark green color, and almost without a flaw. Yet no mines producing emeralds are worked in Bavaria, though it is probable they were known to the Romans. The stones are embedded in the sides of two tall perpendicular rocks; so steep indeed, that access to the treasures they contain is only possible to the adventurous spirit who chooses to lower himself by means of ropes, and remain suspended over a frightful chasm, while he detaches with his tools the emeralds which some cataclysm of nature, cleaving the rock asunder, has brought to light. In the district, this perilous and fatiguing undertaking is called "*abseilen*," from the word "*seil*," rope, or cable. Among the very few persons who have ventured the descent, we have known a woman. This heroine found an ample reward in the number of very fine emeralds she detached during her aërial expedition.

CHAPTER II.

GEOGRAPHY OF DIAMONDS (IN THE OLD WORLD).

Diamonds, where found.—Travels of Tavernier.—Diamond Mines of India; mode of working them.—Juvenile Merchants.—Singular hiding-place for Stolen Goods.—Silent Mode of Barter.—The Koh-i-Noor as seen by Tavernier.—Diamond Mines of the Island of Borneo, of Siberia, of Algiers.

"Mais j'ai vu scintiller le diamant son frère, (du cristal)
Jadis de son berceau nous cachant le mystère;
Il rayonne à vos doigts, il pare vos cheveux:
Pouvez-vous ignorer la source de ses feux!
Daubenton vous dira quelle arène féconde
Aux champs de Visapour, aux rochers de Golconde,
Dans les flots détrempés et retrempés encore,
Laissa du sable avare échapper le trésor.
Dans ton sein quelquefois l'onde le voit éclore;
Quelquefois des métaux la vapeur le colore.
Et de sa croûte épaisse enlevant les débris,
L'art en le polissant en réhausse le prix.
Les rois, les potentats, ainsi que la victoire,
D'un diamant fameux se disputent la gloire.
Son éclat de leur trône accroit la majesté;
Il pare la grandeur, il orne la beauté.
Et pour comble d'honneur, ce Newton qui des mondes
Dirigea dans les cieux les sphères vagabondes,
Jetant un œil perçant dans l'avenir lointain,
Devina son essence, et predit son destin.
Du choix des éléments, formé par un long âge
Des pouvoirs minéraux le plus parfait ouvrage.
Tant de beautés vaut bien qu'en se parant de lui,
Eglé pour le connaître endure un peu d'ennui."

DELILLE—*Trois Regnes de la Nature.*

DURING the middle ages, the birth-place of gems remained almost as much of a mystery as it had been with the ancients. The Venetian merchants, the first navigators to the East Indies, kept their own counsel as to the precise spot whence they drew their rich freight.

In Arabic and Persian works of natural history, Aristotle is generally quoted as the chief authority whence information is drawn, and the most vague and fabulous tales of the origin and qualities of natural substances are laid to his account; many no doubt with justice, but more without any authority whatever. Thus of the diamond some authors assert, that when Alexander visited the mountain Zulmeah (others call it Sarandip), where the inaccessible valley of diamonds is situated, he directed pieces of flesh to be thrown in as the only means of procuring the gems: vultures picked up these with the precious stones attached to them, and dropped them in their flight on various parts of the earth, where alone they are now discovered*.

Marco Polo, who travelled in India in the early part of the thirteenth century, gives much the same account of the mode of procuring diamonds, as having been given to him by the inhabitants "They" (the inhabitants) "told him that in the summer, when the heat is excessive and there is no rain, they ascend the mountains with great fatigue, as well as with considerable danger, from the number of snakes with which they are infested. Near the summit, it is said, there are deep valleys, full of caverns and surrounded by precipices, amongst which the diamonds are found; and here many eagles and white storks, attracted by the snakes on which they are accustomed to feed, make their nests. The persons who are in quest of the diamonds take their stand near the mouth of the cavern, and from thence cast down several pieces of flesh, which the eagles and storks pursue into the valleys and carry off with them to the tops of the rocks. Thither the men immediately ascend, drive the birds away, and recovering the pieces of meat, frequently find diamonds sticking to them."

This relation of the mode of obtaining precious stones

* Oriental accounts of the precious minerals, translated by Rajah Kalikischen, with remarks by James Prinsep, F.R.S., in the Journal of the Asiatic Society of Bengal. Calcutta, 1832.

from an inaccessible valley, is identical with the story in one of the adventures of Sinbad, the sailor, in the *Arabian Nights*. It is probable that the story of the valley of diamonds was current in the Eastern world, and its antiquity is satisfactorily proved by Epiphanius, in his work *De duodecim lapidibus rationali sacerdotis infixis*, written in the fourth century.

Diamonds are found disseminated in the ferruginous sands that constitute ancient alluviums, and in the beds of rivers. From the time the diamond was first known and appreciated, to the beginning of the eighteenth century, India was supposed to be the only land where it was found. The few travellers, who, in the seventeenth century, enlightened Europeans with regard to the East Indies,—that far-off land, whose name was significant of every luxury, of all that was beautiful, rich, rare, and marvellous,—were merchant-jewellers, and, among these, Jean Baptiste Tavernier takes the first place. In fact, he was the first to publish a detailed and faithful account of the diamond mines, of the manner in which they were worked, and of the trade carried on in their produce.

His travels, though still appealed to as a chief authority in all that regards gems, are yet so little read, that a few extracts from his works, and a slight notice of the author, may be new to many of our readers. This celebrated traveller, whose narratives were, in his own time, the subject of so much praise and so much censure, and whose testimony as a merchant-jeweller and an ocular witness, is doubly valuable, deserves implicit credit whenever he speaks from his own personal observation, for neither his judgment nor his veracity can then be impugned. When, however, Tavernier relates from hearsay, his account is often greatly exaggerated, or totally unfounded. In many instances where his assertions had drawn upon him the ridicule of his contemporaries, they have been confirmed by the testimony of recent travellers.

Jean Baptiste Tavernier was born in 1605, in Paris, where his father, a native of Antwerp, had settled as a dealer in maps. Young Tavernier learned the jeweller's trade, and the knowledge of gems he thus acquired subsequently proved a source of great profit and pleasure to him. The conversation of those who came to make purchases at his father's shop, stimulated the natural inclination of the youth. He commenced, at the age of twenty-two, his travels through France, England, the Netherlands, Germany, Switzerland, Hungary, and Italy. He spent forty years of his life travelling in Turkey, the East Indies, and Persia, where he made a large fortune by trading in precious stones. On his return to France in 1664, he was ennobled by Louis XIV., and purchased the barony of Aubonne, in the Canton of Berne, on the margin of the lake of Geneva. Here, among his co-religionists,—for Tavernier was a Protestant,—he had hoped to spend the remainder of his days in the enjoyment of the wealth he had so laboriously acquired; but the bad management of a nephew, to whom he had confided a large venture, again sent him abroad. A freight, worth two hundred and twenty-two thousand livres, from which he had hoped to obtain in the Levant a profit of more than a million, proved so great a loss, that to repair it, and enable himself to speculate again, he sold his estate, and, at the advanced age of eighty-four years, set out for the East. He probably deemed that his long experience, and consummate knowledge of the trade in the countries he had formerly visited, would soon enable him to retrieve his losses, forgetting that he lacked the elasticity of youth, the strength of maturity. He died on his way, at Moscow, in 1689.

In the six voyages he made in Turkey, Persia, and the East Indies, during nearly half a century, Tavernier, who was a man of acute observation and excellent judgment, acquired a large amount of useful and interesting information. But so long an absence from his native land, and constant intercourse with foreign nations, had incapacitated him to edit his travels, and they were, therefore, put into form and published by Chapuzeau and La Chapelle (3 vols., Paris, 1677-79).

Tavernier visited three diamond mines, that of Raolconda, near Visapoor; that of Coloor, in the Circars, now British property, some thirty leagues west of Masulipatam; and that of Sumelpoor, or Guel, on the south-western frontier of Bengal. The real age of these mines never has been ascertained.

The first diamonds that were known to the European trade, were brought from the kingdoms of Visapoor and Golconda. The discovery of the mine that has given such celebrity to Golconda, is attributed to a poor shepherd, who, while tending his flocks, stumbled upon what appeared to him a pretty pebble: this pebble he bartered to some one as ignorant as himself, for a little rice. After passing thus through several hands, it fell into those of a merchant who knew its worth, and who, after diligent search, succeeded in finding the mine. Having made excavations, he came upon a reddish earth mixed with pebbles, and intersected by white and yellow veins of a substance resembling lime.

Tavernier supposed he was the first European that had visited the mines of Golconda; but he was mistaken. An Englishman of the name of Methold, had been there before him, in 1622, and found thirty thousand laborers at work in the mine he visited, and which he stated to be but two leagues from the capital. It was then farmed to Marcandar, a rich merchant-jeweller, who paid three hundred thousand pagodas to the King annually for the right of working it; all stones weighing more than two carats were also reserved for the King. Marcandar had divided the soil into square lots, which he rented to other merchants. Severe punishments were inflicted on whosoever attempted to defraud the sovereign of his dues; but this did not prevent a quantity of fine diamonds from finding their way into the market without passing through the royal hands. Methold saw two that each weighed nearly twenty carats, and several weighing each ten or twelve carats. Such was the high esteem in which fine stones were held in the country that, notwithstanding the risk of detection, the seller always obtained large prices.

The diamond mines of Golconda are not in the neighbourhood of the fortress, the soil of which has never produced precious stones of any kind. The gems were found at the base of the Neela-Mulla mountains, in the vicinity of the Krischna and Pomar rivers. That district is naturally so sterile that it was probably a desert previous to the discovery of its mineral treasures. When Methold was there, notwithstanding the dearness of provisions, in consequence of the distance from which they were brought, there was a population of a hundred thousand souls, the majority of which were miners, laborers, and merchants. The stones were conveyed to Golconda in a rough state, and there cut and polished. This place becoming the chief mart, it is commonly supposed in Europe that the gems were found close by. The mines formerly furnishing these much-prized stones are now exhausted and abandoned. Even when Methold quitted Masulipatam, the mine he had visited, at four days' journey from that place, was nearly exhausted.

Modern geologists are of opinion that the most productive diamond mines exist in virgin soil, yet untried by the natives of India, who have no scientific data to guide their researches. In fact, it is supposed that the strata of many parts of the country is of diamonds, and that the earth contains inexhaustible treasures of this nature.

The account of the mines of Golconda, given by Tavernier, is much more circumstantial. He had visited the Persian Gulf for the purpose of speculating in pearls, and thence resolved to proceed to Golconda to procure fine diamonds and to sell his pearls, the least of which weighed thirty-four carats, to the King.

The first mine he visited was that of Raolconda, then the most famous, and which he was informed by the inhabitants had been opened some two hundred years before his visit. This mine was about five days' journey from Golconda, and eight or nine from Visapoor. The mode of trading there he gives as follows:—

"A very pretty sight is that presented every morning by the children of the master-miners and of other inhabitants of the district.

"The boys, the eldest of which is not over sixteen or the youngest under ten, assemble and sit under a large tree in the public square of the village. Each has his diamond weight in a bag, hung on one side of his girdle, and on the other a purse containing sometimes as much as five or six hundred pagodas. Here they wait for such persons as have diamonds to sell, either from the vicinity or from any other mine. When a diamond is brought to them, it is immediately handed to the eldest boy, who is tacitly acknowledged as the head of this little band. By him it is carefully examined, and then passed to his neighbour, who having also inspected it, transmits it to the next boy. The stone is thus passed from hand to hand amid unbroken silence, until it returns to that of the eldest, who then asks the price, and makes the bargain. If the little man is thought by his comrades to have given too high a price, he must keep the stone on his own account. In the evening the children take account of stock, examine their purchases, and class them according to their water, size, and purity, putting on each stone the price they expect to get for it; they then carry the stones to the masters, who have always assortments to complete, and the profits are divided among the young traders, with this difference in favor of the head of the firm, that he receives one-fourth per cent. more than the others. These children are so perfectly acquainted with the value of all sorts of gems, that if one of them, after buying a stone, is willing to lose one-half per cent. on it, a companion is always ready to take it."

The laborers in the mines being wretchedly remunerated,—for their wages do not amount to more than three pagodas yearly—never hesitate, if they can get a chance, to conceal a diamond when they find one; this, as their only dress is a piece of cotton cloth round their loins, is no easy matter. Sometimes they swallow it. Tavernier tells of a miner who concealed a stone of two carats in the corner of his eye; he was however detected.

The merchants who repair to the mine to make purchases do not go forth to seek the sellers; they wait at their lodgings, where the master-miner visits them in the morning with samples. If the lots are large, they are left with the merchant a day or two, to allow time for consideration. If, when the time is up, there is any hesitation on the part of the buyer, the masters tie their diamonds up again in a corner of their shirt or girdle and are off; they never return with the same stones, or else they mingle them with others, so as to alter the price previously demanded.

The mode of bargain between these traders was peculiar. The whole transaction took place in the most absolute silence. Buyer and seller seated themselves opposite to each other, cross-legged, on the ground. The seller taking the purchaser's hand, placed it under his own girdle, and under this covering the bargain was carried on and concluded, solely through the medium of the two hands, without the aid of eyes or tongue, and that too with the greatest secrecy, although in the presence of several other merchants. If the seller clasped all the hand of the purchaser, the price was a thousand rupees, or pagodas, according to the coin in which they dealt. So often as the clasp was renewed, so many were the thousands implied. If five fingers only were held, five hundred coins were expressed; one finger meant one hundred; half a finger to the middle joint went for fifty: one joint for ten. It sometimes happened that the same lot was disposed of on the same spot, to various persons in turn, before the same witnesses, seven or eight times, no one but the parties interested knowing each time what the price was. There never could be any cheating in weight, except in clandestine sales, as an officer appointed by the King weighed the diamonds without any charge to the merchants, who were to abide by his decisions.

The next mine visited by Tavernier was that of Gani, called by the Persians Coloor, in the same kingdom, and about seven days' journey from the capital. In his day there were sixty thousand laborers at work there. This mine was also discovered by chance about a century before his visit to it. The finder of the first diamond was a poor man, who was preparing his ground to sow millet. He, however, was aware of the value of his prize, and carried it to the capital, where its size greatly surprised the dealers; it weighed twenty-five carats, and they had not yet met with stones weighing over twelve.

Much larger ones were subsequently found in this mine; stones of forty carats were frequently met with, and Tavernier speaks of a very fine one presented by the Indian general Mirgimola to Cha-Gehan, the father of Aureng-Zebe, after he had betrayed his former master, the King of Golconda. When in its rough state, it weighed seven hundred and eighty-seven and a-half carats; —elsewhere Tavernier speaks of the weight as being seven hundred and ninety-three carats.

Though so fine in point of size, the French jeweller found the stones of this district, in general, inferior in point of purity and color, being tinged with green, yellow, or red.

The third and most important mine visited by Tavernier, and worked in his day, was that of Sumelpoor, on the south-western frontier of Bengal, which derived its name from a large village in the vicinity; it was also called, "Guel," from a river of that name which runs at its foot, and where diamonds were also found.

Besides the mines visited by Tavernier, there are three others: that of Gandicotta, in the ancient territories of Tippoo, about sixty miles north-west of Madras, between Gooti and Cuddalah; that of Beiragoor, thirty leagues south of Sumelpoor; and, lastly, a third situated above the peninsula, near Panna, some thirty leagues south-west of Allahabad, on the Ganges.

The mines of Panna are the most interesting to the antiquary, inasmuch as they prove the existence of diamond districts in the part of India that was known to the ancients. The region of Panna was the land of the Prasians, the most powerful of the Indian nations. Their capital, Patibothra, in the vicinity of Patna, is generally looked upon as the capital of India.

The *Jawahir-nameh* describes the mode of digging for the ore and washing the sand or gravel. The similarity between the diamond and rock crystal, both met with in the same matrix, has given to the latter the appellation of *Kacha*, or the *unripe*, and to the real gem, that of *Pakka*, or *ripe* diamond*.

There are diamond mines in the island of Borneo.

Within the last five-and-twenty years diamond sands have been discovered in Siberia, on the western slope of the Oural mountains, near Keskanar, 250 versts from the town of Peru.

Even Africa is expected to yield diamonds. Three have been found in the auriferous sands of the river Gounel, in the province of Constantine. One of these

* Oriental accounts of the precious minerals, by Rajah Kalikischen, with remarks by James Prinsep, F.R.S., in the Journal of the Asiatic Society of Bengal. Calcutta, 1832.

stones, weighing three carats, and worth about £20 if free from flaw, is now at the École des Mines in Paris; another, which is in the Museum of Natural History, weighs one and a quarter carat; and the third, weighing one carat, passed into the collection of M. Drée.

CHAPTER III.

GEOGRAPHY OF PRECIOUS STONES IN THE OLD AND NEW WORLDS.

Ruby and Sapphire Mines of Pegu and Ava.—Monopoly of Rubies by the Indian Princes.—The Island of Ceylon and its Gems.—Difficulty of ascertaining where the Ancients procured their Gems.—Ancient Authorities: Ctesias, Theophrastus, Herodotus.—Trade carried on by the Phœnicians and Carthaginians in Precious and Fine Stones.—Influence of the Conquests of Mexico and Peru on the Commerce of Gems.—Immense quantities of Emeralds brought from the New World by the Companions of Cortez and Pizarro.—Phenomenal Emerald.—The five Emeralds of Cortez.—Trying Test.—Demon-guarded Emerald Mines of Peru.—Revival of the Emerald Trade of late years.

ALMOST all the hyaline corindons are brought from Pegu, Ava, and the island of Ceylon. Tavernier says, that in the kingdom of Pegu they are found in a mountain called Capillan, some twelve days' journey north-east from the town of Siren, where the King resides. From this mountain are drawn rubies, yellow topazes, blue and white sapphires, amethysts, etc. He adds, that it was impossible to go to the mines by land on account of the immense number of wild beasts that infested the forests that were to be traversed.

The high value of the ruby proceeds not only from the rarity of perfect stones, but also from the care taken by the Indian princes, with whom this gem is an especial favorite, to prevent the fine ones from being taken out of the kingdom. Indeed, they are always ready to give higher prices for extraordinary gems than can be had in Europe. This was also the case in Tavernier's day: he tells us, it was almost impossible to procure fine rubies of three or more carats; and that he found more profit in bringing such stones back from Europe, than in taking them there from Asia. The rubies of Ceylon are esteemed the best; they are brought down from the high mountains, by the torrents, into the beds of the streams.

The following account of the ruby and sapphire mines of Pegu and Ava, is from the pen of Mr. Crawfurd, a late traveller:—"The precious stones ascertained to exist in the Burmese territory are chiefly those of the sapphire family and the spinel ruby. They are found at two places not very distant from each other, called Mogaut and Kyatpean, about five days' journey from the capital in an E.S.E. direction. From what I could learn the gems are not obtained by any regular mining operations, but by digging and washing the gravel in the beds of rivulets or small brooks. All the varieties of the sapphire, as well as the spinel, are found together, and along with them large quantities of corundum. The varieties ascertained to exist are, the oriental sapphire, the oriental ruby r red stone, the opalescent ruby, or cat's-eye ruby, the star ruby, the green, the yellow, and the white sapphires, and the oriental amethyst. The common sapphire is by far the most frequent, but, in comparison with the ruby, is very little prized by the Burmese, in which they agree with other nations. I brought with me several of great size, the largest weighing no less than 3,630 grains, or about 907 carats. The spinel ruby (zebu-gaong) is not unfrequent in Ava, but is not much valued by the natives. I brought with me to England a perfect specimen, both as to color and freedom from flaws, weighing twenty-two carats. The sapphire and ruby mines are considered the property of the King; at least he lays claim to all stones that exceed in value a viss of silver, or two ticals. The miners, it appears, endeavour to evade this by breaking the large stones into fragments. In the royal treasury there are, notwithstanding, many fine stones of both descriptions. The year before our visit, the King received from the mines a ruby weighing one hundred and twenty-four grains; and the year preceding that, eight good ones, but of smaller size. No stranger is permitted to visit the mines; even the Chinese and Mahommedans residing at Ava, are carefully excluded*."

When Pegu was annexed to the British possessions in 1862, hopes were entertained, that its richest products, hitherto so jealously guarded, would find their way to Europe. But no difference has, as yet, been found either in size or quantity. This part of Asia is, however, but very little known, and the accounts given by traders of the number of tigers, lions, and venomous serpents, that infest the interior, whether true, or exaggerated to enhance the value of gems procured amid such dangers, are sufficient to deter the generality of travellers from attempting to explore these dangerous regions.

The finest colored gems are supposed to be monopolized by the Asiatic princes, and the Subbah of Deccan was in possession of a particularly fine ruby, a full inch in diameter, while the finest in the world was in the possession of the King of Pegu; its purity has passed into a proverb, and its worth, when compared with gold, is inestimable.

Among European princes, there are few rubies of any magnitude, compared with those belonging to Indian potentates; but Tavernier gave the preference for beauty to the rubies of the King of France, over any of those of the Great Mogul. Yet the French traveller tells us, he counted on the Indian monarch's throne, one hundred and eight large rubies, varying in weight from one hundred to two hundred

* *Journal of an Embassy to the Court of Ava*, p. 442.

carats, and one single gem, that weighed two ounces and a-half! They were, however, balass rubies.

The *oriental beryls* come from the East Indies, the borders of the Euphrates, and the foot of Mount Taurus.

The oriental chrysolite, which is very rare, is only found in the island of Ceylon.

The finest garnets come from Syria, Calicut, Cambye, and Cananor. An oriental garnet of an orange red hue, drawing on a hyacinthine yellow, very hard, and very rich in color, comes from a town of Pegu called *Sorian*, for which reason this stone was called *soranus* by the ancients. Garnets are found in slates and other foliated and talcous stones, also in lime stones. They re sometimes met with detached, on mountains, and in the sands of rivers.

The finest of the western garnets are from mines in Hungary and Bohemia. Though considered greatly inferior to the oriental, some specimens are exceedingly beautiful. The following passage from Tavernier proves, that even this great connoisseur actually mistook Bohemian garnets for rubies:—"In Bohemia, there are mines that produce pebbles of various sizes, some as large as eggs, others as large as one's fist. When broken, some of these are found to contain rubies as hard and fine as those of Pegu. I remember being one day at Prague, with the Viceroy of Hungary, in whose service I then was, when he and General Wallenstein, Duke of Friedland, were washing their hands before sitting down to dinner. The Viceroy noticed, and greatly praised, the beauty of a ruby the General wore in a ring, but his admiration was increased when he was told, that the mine whence it came was in Bohemia. When the Viceroy departed, the Duke presented him with a basket containing a hundred of the pebbles. On his arrival home the Viceroy had the stones broken; but out of the hundred, only two were found to contain a ruby; one a large gem, weighing five carats; the other a ruby of one carat."

It is very difficult to ascertain, with any degree of precision, whence the ancients drew their gems. The word *topaz* derived from an island in the Red Sea, where, it is supposed, the ancients used to find that gem, was, however, applied by them to a mineral very different from ours. Our variety of the topaz was denominated by them *Chrysolite.*

Ctesias tells us expressly, that the fine stones of which the Babylonians made seals, came from India; and that the sardonyxes, onyxes, and other stones for signets, were picked up on the mountains adjoining the desert. The testimony of modern travellers has confirmed this author's report; and to this day, the finest lapis-lazuli is still found in those parts.

If the assertions of Ctesias are to be credited, we may take it for granted, that the north of India is the original soil whence gems were extracted. The proof that the sapphire of the ancients, called by us *lapis-lazuli,* had its birth there, is furnished by a rather more modern and very reliable authority.

"The emeralds and jaspers," says Theophrastus, "that are used as ornaments, are brought from the desert of Bactriana, or Cobi. These stones are picked up by the riders who go there in the season of the north winds, when these winds raise and carry off the sands." And in another place, he says: "The largest emerald, of those called Bactrians, is found in Tyre, in the temple of Hercules. It is a column of rather large dimensions." This was probably the same pillar of which Herodotus speaks: "In the temple of the Tyrian Hercules," says the historian, "I saw two pillars, one of massive gold, the other of emeralds, gleaming in the dark." Heeren conjectures this pillar to have been of lapis-lazuli. like that in the church of the Jesuits at Rome.

Alluding to the foregoing passage from Ctesias, Heeran remarks, that it contains indications which, as regard onyxes, would seem to designate the mountains of Gate as their original soil, as that passage speaks of a burning soil, and a neighbouring sea. And a farther proof of his is the fact, that onyxes are, to the present day, found in large quantities on the same spot, that is, on the mountains situated near Cambaya and Beroach, which is the ancient Barygasa;—a proof the more decisive, as that part of the coast of Deccan was just that which the ancients were best acquainted with, and that their navigation from the Persian Gulf to those regions, is a well authenticated fact.

Six hundred years before the Christian era, we find the Greeks and the Phœnicians, the pioneers of civilization. At that remote period, the Tyrian Hercules was sailing through every sea, buying and transporting from each country its most precious productions. He did not overlook the fine garnets of the coast of Gaul, or the coral of the Hieres, and the produce of the precious mines which were then worked upon the surface of the Cevennes, the Pyrenees, and the Alps. Penetrating into the Baltic Sea, he brought thence the amber, that then, as now, enriched its margins. The Carthaginians inherited from the Phœnicians their genius for trade, and became the possessors of the lands from which their most precious merchandise had been brought. From Spain's rich mines came the gold, that furnished the legions, that enabled them to resist the power of Rome, and supported their ambitious and factious chiefs. From the interior of Africa they also brought gold as well as precious stones, and the black slaves much sought for in Italy and Greece.

The fine stone then called chalcedonyx, and which drew its name from Carthage, was brought from the mountains of the interior of Africa, to the country of the Garamantes, where the Carthaginians went for them. This stone then also called, as it sometimes is even to the present day, *carbuncle*, held the first rank among the onyxes; vases and cups were made of it, and the immense quantity of these used by the wealthy, may give an idea of the extent of the trade carried on in chalcedonyxes.

Sardinia was one of the most valuable possessions of this commercial nation, not only on account of its geographical position, but also for its mines of gold, silver, and fine stones, and its agricultural fertility. The last advantage, if we may credit an ancient author, seems to have been despised by the Carthaginians, who ordered all the fruit trees on the island to be destroyed, and prohibited, under pain of death, the cultivation of the soil. This strange prohibition, coming from a people who understood so well the value of a fertile soil, can only be explained by the superior importance in which they held its mineral wealth, and the desire to confine the attention of its inhabitants to the extraction of this great article of trade.

One of the chief gem-producing countries of the East is Ceylon, which was known to the earliest historians under the name of Taprobane; the first account of the island was brought to Europe by the Macedonians, who were with Alexander in India. Taprobane contributed largely to supply the demands of the luxury that marked the decline of the sometime mistress of the world. The silks of China, the precious stones of Ceylon, and the rich spices and aromatics of India, were the articles of trade principally sought by the Roman navigators. In exchange for these costly articles, Rome was compelled to pay with her silver and gold, and of the £800,000 she thus annually spent, a large proportion went to the capital of Ceylon. Of the prosperity and commercial importance that was the result to that capital (Anarajahpoora) we may judge, when we hear, that its perimeter exceeded sixty miles. After the ruin of the Roman Colossus, Ceylon rapidly declined, and, from being the chief emporium of commerce in the East, again merged into the barbarism from which contact with the elegant Persians and Romans had partly raised it. Domestic tumults, intestinal broils, famine, and the sword, at last reduced the island to the degenerate state in which it was found by the Portuguese in the beginning of the sixteenth century.

Ceylon seems to experience no decrease in her rich mineral productions, and quantities of gems are still brought from thence. The most admired are the ruby, the sapphire, and the amethyst. The ruby is found there very fine; though those of Pegu are said to be finer still. The sapphire is the most abundant of the hyaline corindons, and the blue is sometimes met with of very large size. The black sapphire is uncommon, and, when procured, is in very small specimens. One of the most admired gems found in Ceylon is the *asteria*, or star-stone, a very beautiful and singular variety of the sapphire of a greyish blue color, which, when subjected to a strong light, presents a star composed of six delicate white rays, turn it whatever way you may. The finest cat's-eyes in the world are those procured in this island, and indeed the only ones that bring high prices. The native topaz commonly passes under the name of "the white and water sapphire." It is generally white, bluish, and yellow, and commonly much deteriorated by attrition; perfect crystals of it are exceedingly rare. The purple variety of the amethyst is rare, but the green still more so. The common garnet is very abundant, but its crystals are small, and very apt to decompose. The precious garnet is rare; and when found, is not of good quality. Cinnamon stone, a species of garnet of little value, is peculiar to this island. It is sometimes found in large masses, though more frequently in small irregular fragments.

Rock crystal occurs in abundance, both massive and crystallized, of various colors and in large masses. The gem known by the name of the *matura* diamond is nothing more than a fine crystal, yet is rather prized by the more wealthy natives. The finest pearls are procured from the oyster beds on the coast. The Cingalese, like all East Indians, are extremely fond of gems, and the rich spend incredible sums in purchasing the rarest; the result is, that the worst only find their way into the foreign markets.

From a modern traveller we borrow the following interesting account of the jewellers, or rather lapidaries, of Ceylon: "They sit under a verandah or shed, in front of the house, squatted on their heels, behind a rude lathe, raised a few inches from the ground. On the end of its axle is a round plate of iron or steel about eight inches in diameter, placed vertically; which is made to revolve backwards and forwards by a drill-bow about four inches long, made of bamboo, and worked by the right hand, while the left applies the stone to be cut, held tightly between the finger and thumb against the wheel. A sort of emery or finely-powdered sapphire of coarse quality, moistened with water, is the only intermediate substance used in cutting the stone. One of the lapidaries, who seemed to be indifferently honest, told me that what are called "Ceylon diamonds," are made of a species of tourmalin, which is boiled for some time in cocoa-nut oil, before being cut, to make it perfectly transparent. A gentleman from the ship saw one of these jewellers manufacturing white sapphire, from the fragments of a decanter and a glass fruit-bowl*."

The principal mines of oriental agates at the present day are in the little principality of Rajpepla, in the province of Gujerat, fourteen miles distant from the town of Broach, where they are cut into different ornaments and trinkets. They are exported in great quantities to other parts of India and to Europe, and hence perhaps the jeweller's term "brooch."

The oriental cornelian comes from the East Indies, Arabia Petrea, Persia, and Egypt.

* *Narrative of a Voyage Round the World*, by W. S. W. Ruschenberger, M.D.

The oriental girasol, or argentine, is found in Asia Minor.

The oriental, or spangled opal, is brought from Ceylon. Black opals, which are very rare, are found in Egypt.

The genuine turkois is the produce of the mines of *Ansar*, near Nishapoor, in Khorassan (the same place mentioned as Nichebourg in Tavernier's Travels). All authorities concur that these are the only turkois mines in the world: the stones vary from pale blue to green and white, but all except the azure are worthless. A curious fact is mentioned also, which, from the nature of the mineral, may be readily believed, though it has not been observed in Europe; the real blue turkois of Nishapoor changes its color when kept near musk or camphor, also from dampness of the ground, as well as from exposure to the fire; the inferior stones become discolored even without this test, by gradual decomposition or efflorescence. It is found in pieces of a semi-orbicular form, the largest the size of nuts, embedded in a ferruginous clay that fills the fissures, in a soil the nature of which is little known, but which appears to belong to silicious schistus. Large fragments are seldom pure. In Moscow it is sold in lots of about a thousand fragments, the size of peas, for five roubles (1*l.*) the lot.

When the conquest of Mexico and Peru turned the tide of commerce from the East to the West, on no article of trade was the influence of these great events so visible as on gems. It was with no less consternation than surprise that the dealers viewed continual arrivals of precious stones in fabulous quantities, and surpassing in size and beauty all they had hitherto known. That they had cause for both alarm and wonder, the narratives of contemporaries sufficiently prove. Pierre de Rosnel, in his *Mercure Indien, ou le Trésor des Indes*, published in 1664, speaking of the immense quantity of emeralds then drawn from Peru, relates that a certain Spaniard who had spent some time in Cuzco, being in Italy, showed an emerald to a lapidary and requested to know its value. The lapidary valued it at a hundred ducats: upon which the Spaniard showed him a second larger one, which was valued at thrice the amount. Much astonished at this difference, and hoping to sell all he had in accordance with the prices named, he invited the lapidary home, and showed him a large chest full of emeralds, which, taken one with the other, might have been worth fifty ducats each. The lapidary, however, seeing such a quantity, and knowing that such things are valuable only from their scarcity, immediately changed his tone, and offered one crown apiece for the stones.

In corroboration of the belief that formerly existed with regard to this stone being *born white* and ripening into its beautiful green color during its residence in the mine, the same author gives an account of a singular emerald which a Peruvian asserted he had seen in the province of Cuzco. Two corners of the gem were a beautiful green, the remainder was a pure white. The King of Spain, Philip II., expressed a great desire to possess this phenomenon; but unfortunately its owner, under the mistaken idea that this singularity detracted from the merit of the stone, had caused the white corners to be rounded off.

In the booty obtained by Fernando Cortez from the province called the Golden Castile, were five emeralds then valued at 100,000 crowns. The first was cut in the shape of a rose, with its leaves; the second was a toy; the third a fish; the fourth was a bell, the clapper of which was a large pear-shaped pearl; the fifth was a cup, for which a Genoese lapidary offered 40,000 ducats.

An immense number of emeralds fell into the hands of the conquerors of Peru, and many were exceedingly fine; one is mentioned as being of the size of a pigeon's egg. When the temple of the Sun was pillaged, a vast number were found there. When Atahuapla, the Peruvian Inca, fell into the hands of the Spaniards, he wore a collar of emeralds of very large size and great brilliancy.

Unfortunately one of the Dominican monks who accompanied the expedition, persuaded the soldiers that emeralds could not be broken, and thus a quantity of these beautiful gems were wantonly destroyed in the attempt to test, by means of hammer and anvil, their genuineness. The adviser did not probably subject his own jewels to this trying process; it was even suspected that his counsel was not altogether disinterested, as, in consequence of this depreciation of the stones, he was enabled to obtain a large number for little or nothing.

For the last two centuries and more the only country known to yield emeralds has been Peru, where they occur in Santa Fé and the valley of Junca. In the province of Quito was the so celebrated river of Emeralds, which drew its name from the quarries of those gems on its margins, whence the Indian monarchs obtained their jewels.

A singular analogy appears to exist between the superstitions of the old and the new world, in past and present times. The belief that demons, griffins, or wicked spirits guard the treasures contained in emerald mines, is as strong at the present day, among the Peruvian Indians, as it was in the age of Pliny with regard to the Scythian mines, or as it is now with the Arabs in the vicinity of Mount Zebarah. Stevenson, speaking of the emerald mine in the neighbourhood of "Las Esmeraldas," says: "I never visited it owing to the superstitious dread of the natives, who assured me that it was enchanted, and guarded by an enchanted dragon, which poured forth thunder and lightning on those who dared to ascend the river*!"

For some time the Peruvian emerald mines were supposed to be exhausted; about twenty years ago, however,

* *Residence in South America*, vol. ii., p. 406.

Mr. Mention, head of a large Parisian firm, received from South America some splendid specimens, which revived the emerald trade, continued since then without interruption by Mr. Charles Achard. The emerald mines from which the ancient Mexicans and Peruvians had their phenomenal gems, are lost to us, however. It is conjectured that they drew them from the section of the Cordilleras, which, running from north-east to south-west, separated the unexplored land of the Candones (Lacandones, or Caribes) from the province of Chiapas. When Las Casas attempted to convert the Candones, they received the venerable man with deference, and the few Spaniards who accompanied him purchased of them the finest emeralds in the rough, or rudely polished, that have ever come to Europe, —such as, indeed, have never found their way there since.

Fine turkoises are found in the provinces of Guanaxuato, in the north-western part of Mexico, and in the province of Oaxaca, in the south-west.

Fine opals are also found in Mexico, in the mines of Guanaxuato, Oaxaca, Zacatecas, Xalisco, Aguas Calientes, Anganguea, etc.

CHAPTER IV.

GEOGRAPHY OF DIAMONDS IN THE NEW WORLD.

Discovery of the Diamond Mines of Brazil.—Diamond Districts of Serro do Frio.—Diamond Mines of the Province of Matto Grosso.—Process of Extraction.—Former Prejudice against Brazilian Diamonds.—Fallacy of Judgment of great Jewellers in the past Century.—Brobdignagian Jewel.—All that glitters not a Diamond.

THE discovery of the diamond mines of Brazil, which occurred in the year 1730, was, like that of the East Indian mines, the result of fortuitous circumstances. Shortly after the establishment of Villa do Principe, the miners searching for gold in the rivulets of Milho Verde and San Gonzalez, in the district of Serro do Frio, met with some singular pebbles of peculiar hue and lustre, which they carried home to their masters as curiosities. Considered merely as pretty baubles, the stones were given to children and used as counters. They at last attracted the attention of an officer who had spent some years at Goa, in the East Indies. Struck with their geometrical symmetry and their weight, he weighed one of them against a common pebble of equal size, and found the counter much the heaviest. Having rubbed the counter on a stone with water, he could make no impression whatever on the former, while a flat surface was easily produced on a common pebble by the same process in a few minutes. He then sent a handful of the singular stones to a friend in Lisbon, with a request that he would have them examined. But the lapidaries of that city, who never wrought diamonds, and probably had never seen one in its rough state, replied that their tools could make no impression upon them. The Dutch Consul, however, chanced to see the counters, and declared it his opinion that they were diamonds. A few were immediately collected and sent to Holland, where they were cut as brilliants and pronounced, by the astonished lapidaries, to be equal to Golconda diamonds of the first water. The news soon reached Brazil, and fortunate were the individuals possessed of some of these hitherto little esteemed counters. Numbers that were scattered in as many hands, were in three or four days bought up by a few individuals. A decree was now issued by the Portuguese government declaring all diamonds found in its soil a monopoly of the crown, and search was instituted in good earnest for the sparkling gems.

The diamond district of the Serro do Frio is about twenty leagues in length, and nine in breadth; the soil is barren, but intersected by numerous streams. It was long supposed that diamonds were confined solely to this district. This was an error; they have been found in every part of the empire, particularly in the remote provinces of Goyazes and Matto Grosso, where there are several diamond localities. These gems have even been found on the tops of the highest mountains; indeed, it is thought by Brazilian mineralogists that the original diamond formations are in the mountains, and that they will one day or other be discovered in such quantities as to render them objects of comparatively small value.

From a late work* we borrow the following interesting particulars on the diamond mines of Brazil:—" The gold, and more especially the diamond mines to which the city of Diamantino is indebted for its origin, appear to have been known to the first Pauliste establishments in the pro-

* *Expédition dans les Parties Centrales de l'Amérique du Sud,* par Francis Castlenau.

vince of Matto Grosso; but under the Portuguese government, the search for diamonds was strictly prohibited to private individuals under the severest penalties. A military guard occupied the diamond district, and watched the slaves of the crown employed in the extraction of this precious mineral. Every individual who found a stone was bound to remit it to the Superintendency of Diamonds of Cuyaba, and in return received a moderate reward, but was punished severely if he was discovered to attempt its appropriation. The diamond trade was as severely prohibited as the extraction, in all the extent of Brazil, to all who were not special agents appointed by government. From the time of Governor Joao Carlos, the trade was gradually more and more tolerated, until it became free altogether. The laws that prohibited it may not have been abrogated, but they have become obsolete. The inhabitants, however, complain that they reap no benefit from the toleration, the interdiction of the slave-trade rendering it impossible to turn to account the wealth of the soil.

"Gold and diamonds, which in these regions, as in many others, are always found united, are gathered more especially in the numerous streams that traverse it, and even in all the extent of its soil. After the rains the children of Diamantino seek for gold in the soil of the streets and in the river Ouro that runs through it; and it is not rare that they pick up the value of eight to fifteen grains. As to diamonds, a negro is reported to have found one of nine carats among the roots of some vegetables which he pulled up in his garden. Diamonds have been sometimes found in the crops of chickens.

"The extraction of these precious gems is carried on in a very simple manner. In the season of low waters the negroes plunge and bring up from the bed of the stream the *cascalho*, or alluvial soil, which is removed to a convenient spot on the banks for working. The process is as follows: a rancho or hut is erected about a hundred feet long, and half that distance in width; down the middle of the area is conveyed a canal covered with earth; on the other side of the area is a flooring of planks, about sixteen feet in width, extending the whole length of the shed, and to which an inclined direction is given; this flooring is divided into troughs, in which is thrown a portion of the cascalho; the water is then let in, and the earth raked until the water becomes clear. The earthy particles having been washed away, the gravel is raked up to the extremity of the trough; the largest stones are thrown out, and afterwards the smaller ones: the whole is then examined with great care for diamonds. When a negro finds one, he claps his hands, and stands up, holding the diamond between his finger and thumb; it is received by one of the overseers posted on lofty seats, at equal distances, along the line of the work. On the conclusion of the work the diamonds found during the day are registered by the head overseer. If a negro has the good fortune to find a stone weighing upwards of seventeen carats, he is immediately manumitted, and for smaller stones proportionate premiums are given.

"The labor of bringing up the *cascalho* can only be done by negroes, as whites could not resist this sort of work in that climate.

"The rivers Diamantino, Ouro, and Paraguay, appear already completely exhausted. The little river Burete still furnishes many stones, but the Santa-Ana seems as yet untouched, so little has it lost from its original wealth, notwithstanding the incredible quantity of diamonds already taken thence. However, it is probable the working is not so profitable as it may seem, since the result obtained by a Spaniard of the name of Don Simon is quoted in the neighbourhood as something very remarkable. In four years, working it is true only during the dry season of each year, but assisted by 200 slaves, he gathered 400 oitavas of diamonds (about 700 carats). He was compelled to give up his search, having lost many of his men by the pestilential fevers which reign throughout the diamond district, but especially on the borders of the Santa-Ana; but before he left, he filled up the cavity whence he had taken his stones.

"Some time afterwards another person found 80 oitavas on one sole point of the river. The largest diamond taken out of the Santa-Ana is said to have weighed 3 oitavas (about 52 carats), but this was some years ago. It is said that the stones taken out of this river are finer than those of any other locality, and that some persons engaged in the trade are able to distinguish them.

"We have said that the negroes employed at this work are strictly watched, and severely punished when detected purloining; but, for all this, they steal a fair half of the produce. On Sundays, and on the numerous fête days, they are permitted to work on their own account: there is always gold in the diamond cascalho, but this their masters will not permit of their wasting their time in taking out. Some negroes, however, do it on their own account, and find it profitable."

When diamonds were discovered in Brazil, the Portuguese fleet brought, in one year, over seventy pounds of diamonds from that country to Europe. Hitherto it was supposed these precious gems only existed in the East Indies. A panic seized the large diamond merchants, who feared their stock in hand would greatly diminish in value in consequence of such an increase, and they took care to spread the most unfavorable reports respecting the new stones.

The attempt to keep up the prices of the Eastern gems by depreciating the South American, was long successful; the latter were deemed very inferior, especially in hardness, and were rather derogatively designated as Portuguese or Brazilian diamonds. Although the new stones were proved fully equal in every respect to the East Indian, the prejudice was so strong against them at one time, that. in order to obtain a fair price for their stones, the Brazilian merchants were in the habit of sending them in the first instance to Goa, that they might be re-imported from that place into Europe, as the productions of the East. The truth, however, at last triumphed, and the Brazilian gems took their place as high in honor as those of the Old World.

In Mr. Mawe's travels in the interior of Brazil, an amusing incident is given, proving that even as all is not gold that glitters, neither is every brilliant a diamond.

A free negro of Villa Rica,—which might more properly be named Poor Town, notwithstanding the gold and diamonds by which it is surrounded,—wrote to the Prince Regent that he possessed a diamond so enormous that he begged to be allowed to present it in person to the Prince himself. A carriage and an escort were forthwith dispatched to take him to court. Blackey threw himself at the Regent's feet, and exhibited his diamond. The prince uttered an exclamation of surpise—the lords present were astounded; the stone weighed nearly a pound! The courtiers immediately set to work to find out the number of millions this monstrous jewel was worth. Mr. Mawe does not give the estimate, but Mr. Hoffman, in his elegant literary criticism of the book, does. The pound contains 9,216 grains, which make 2,304 carats, and even more, for the weight used for the carat of the diamond is lighter than the weight of the marc. The great stone of Villa Rica was valued at Troy weight, which is one-ninth part heavier, consequently 256 are to be added, which, united to the 2,304, make a total of 2,560 carats. Deducting the 60 carats for what little the stone lacked of a pound, there yet remain 2,500 carats. In order to ascertain the commercial value of the stone, the carat must be multiplied by the square. The square of 2,500 is 6,250,000, and estimating the carat at only 150 francs, the common price, we have the sum of 937,500,000 francs; and, as large diamonds are no longer submitted to the tariff, and as their nominal price increases in proportion as they exceed the ordinary dimensions, the Portuguese noblemen probably estimated the stone at two milliards, or, like thorough courtiers, at four.

However this may be, the inestimable jewel was sent to the treasury, with a strong escort, and deposited in the hall of gems. As Mr. Mawe was at Rio Janeiro when this wonderful discovery was made, the minister sent for him, and communicated to him all the particulars regarding the phenomenon; but at the same time, expressed his private doubts of its reality. The English mineralogist was invited to examine the incomparable brilliant, and fix its value. Furnished with a letter from each minister,—without which formality he could not be admitted,—Mr. Mawe went through several rooms, and crossed a great hall hung with crimson and gold, in which was a statue of natural size representing Justice with her scales. Finally he reached a room in which were several chests; three officers, each having a key, opened one of these chests, and the treasurer with much solemnity exhibited the supposed diamond. Before touching the stone Mr. Mawe had already seen that it was nothing but a piece of rounded crystal; he proved this on the instant by *scratching* it with a real diamond, and this luckless scratch at once annihilated all the millions supposed to have been added to the treasury. The Prince Regent received the news very philosophically, losing the imaginary millions very nobly; but poor Blackey, who had come in a carriage, was left to travel back on foot.

The critic adds the following pithy reflection: "It may seem very strange, that in the very birth-place of the diamond, that stone should be so little known, that a piece of rock crystal has thus been allowed to usurp its honors; let us not judge hastily however; we shall see that error is common to all men, and that merchants themselves may be mistaken as well as princes."

PART THIRD.

THE CHEMISTRY OF DIAMONDS AND PRECIOUS STONES.

CHAPTER I.

SCIENCE AMONG THE ANCIENTS AND MODERNS.

Opinions of Ancient Authors as to the Origin of Gems.—Aristoteles.—Theophrastus.—Avicenna.—Falopius.—Cardan.—De Clave.—Boetius de Boot.

THE DIAMOND.

Its Hardness.—Lustre.—Refraction.—Phosphoric and Electric Properties.—Crystallization.—Mode of Testing.—Specific Gravity.—Rough Diamonds.—Mode of Cutting.—The Brilliant, the Rose, the Table, the Lasque.—Different Colors.—Manner of Weighing.—The Carat.—Use made of Poor Diamonds in the Arts.—Commercial Value of Diamonds.—Preference given to them over Colored Hyalines.—Superior Beauty of Diamonds when seen by Artificial Light.—Capital invested in Diamonds not subject to depreciation.—Chemical Nature of the Diamond.—Its Combustibility.—Interesting Experiments.—Attempts to make Diamonds.

"· · · · · · human pride
Is skilful to invent most serious names
To hide its ignorance."—Shelley.

"Oh! que le temps sait bien dans sa marche féconde
Sous mille aspects nouveaux reproduire le monde!
Qui l'eût cru qu'un amas de legers sédiments
Brillerait en cristaux, luirait en diamants!
Que la terre oubliant sa vertu végétale,
Des sucs dus à la fleur colorerait l'opale!
Qu'un ver emprisonné formerait le corail!"—Delille.

In the history of the progress of the human mind we are incessantly reminded of the fact, that men usually reach the truth through the channel of error. Very curious would be the study of all the false systems, the absurd ideas that have preceded what we now look upon as full and true revelations of the mysteries of nature, and which, perhaps, in less than another century, may, in their turn, be dismissed to take their place among obsolete and exploded errors.

At the present day, when science boasts of having disclosed the most hidden of nature's secrets, and raised the last veil; when, arrogantly assimilating her might to that of the Deity, she assumes the power of assigning to each substance its elementary and constitutive principle; we look back wondering and amused at the series of groundless conjectures, extravagant hypotheses, and hap-hazard guesses, through which she has won her way to that which she would now have us accept as the last attainable limits.

The origin of precious stones, their formation, their components, seem to have quite puzzled the brains of the philosophers of old. None, however, would avow his ignorance: each solved the question very arbitrarily, finding some more or less plausible explanation for what was really beyond his comprehension, and publishing his scientific sophisms, his fanciful conjectures, as infallible truths.

The ancient savans were agreed, however, to consider all composite substances to consist of the four elements—fire, air, earth, and water. The difference between gems and common pebbles was supposed to consist in the greater proportion of water contained in the former, while in the latter the earth was predominant. The water, condensed and congealed by the dryness of the earthy particles, was supposed to give to precious stones their lustre and transparency.

Aristotle was the first who undertook to explain the origin and essence of stones in general. He asserted that the first cause of stones was a *viscous mud*, a *sap* that shrinks and is congealed by cold, in which water predominates over earth: to these he adds *stony fragments* and *particles of stone*, and another special *sap*, which he calls *lapidific sap*. As to stones which are neither liquefiable, fusible, or dissolvable, he attributes their formation to a terrestrial, dry, burnt, and igneous exhalation.

Theophrastus. the celebrated author of the *Characters*, and a disciple of Aristotle, but whose literary works bear the stamp of a far more accurate spirit of observation than his scientific writings, merely commented on the opinions of his master. He taught that "stones were made of a pure, equal, and compact substance after its percolation; the which substance, being in certain places, became petrified by flowing, or in some other manner . . . . ."

Avicenna only says that stones are composed of mud, or foul, stagnant water.

Falopius, as regards precious stones, seems to have had a greater predilection for saps than for mud. He asserts them to be the produce of a very pure juice, and not that of a dry exhalation, as the Aristotelians would have it. According to him, "there is no substance in nature so pure as the said sap, with the exception of the natural and vital spirits of animals."

Cardan goes beyond Falopius. According to him, "precious stones are engendered between the rocks by means of a sap, which is distilled through their concavity, even as the babe in the maternal blood." He goes farther still, and explains the variety of color and quality of the different precious stones by that of the different degrees of calcination which nature has given to the distilled juice of which she composes them.

The learned of that day could not stop short on so fair a road. From the allusion made by Cardan *à propos* of the generation of gems in the bosom of rocks, to that of the infant in its mother, a new system was established, the aim of which was to establish that the engendering of precious stones was not to be attributed to the same causes as that of inert nature; but that it proceeded from causes absolutely similar to those of animated living nature.

This system was developed by De Clave, a physician of some celebrity, who, in 1635, published a book, dedicated to Monsieur Séquier, the Garde des Sceaux. The subtleties of reasoning, of allusions, and of language, to which the author was obliged to have recourse in order to give an appearance of truth to his opinion, are most curious and amusing. They must, however, be sought in his book, as it is only in his own quaint, antiquated language one can read what, according to the author, constitute the paternal and maternal organs of this singular generation, and appreciate the nature of the functions attributed to each.

A great step was accomplished, certes, when precious stones were made to pass from the mineral to the animal kingdom; but Boetius de Boot, the physician to the Emperor Rhodolphe II., in a very curious and voluminous work, which he published about the same time, on this subject, went still farther, and attributed the formation of precious stones to the almost direct intervention of the Deity.

After mentioning the opinions of Aristoteles, his *gluey mud* and his *lapidific sap*, Boetius adds:

"Although these things are supposed to be the foundation, yet they do not sufficiently explain to us the substance; inasmuch as what *gluey mud* and *lapidific sap* are composed of, doth not appear. My opinion is, that the first element composing precious stones is an *earth*, so *fine*, *subtle*, and *delicate*, that, being mingled with water, it doth not diminish its transparency; and that the element of common stones is a *coarser*, *fouler earth*. Moreover, there is a *salt* diffused and spread through the bowels of the earth, and also a *fatty exhalation*, which mingles in the composition of both precious and common stones."

Certes, the author of such a theory had little right to

take Aristoteles to task for his gluey mud and lapidific sap. Not content, however, with his *subtle earth*, his *salt*, and his *fatty exhalation*, Boetius tells us all these ingredients would be of no use without the concurrence of the *efficient cause*.

What is discernible through the cloudy metaphysical jargon, by means of which the learned physician seeks to explain the effects of this *efficient cause*, is, that the latter is the spiritual and divine element which animates the substantial and material element of precious stones, the which, if left to itself, would remain inert and without action.

From such starting points did modern chemical science set out. We will now see what she has reached to in 1860.

*　　　*　　　*　　　*　　　*

Among the precious crystallized stones,—the hyaline corindons, as they are termed by mineralogists,—the most valuable in the order in which they are classed by lapidaries, are the diamond, the ruby, the sapphire, the topaz, the emerald, the amethyst, the aquamarine, or beryl, the chrysolite, the garnet, and the hyacinth.

The diamond is the hardest body known; hence the name of adamas, *i.e.* the *indomitable*, given to it by the ancients. It resists the action of the file, and can only be cut and polished by its own dust. This important feature would, in itself, suffice to distinguish it from every other mineral, independently of its other characteristics.

It is also the most brilliant of stones; usually transparent, it is sometimes only semi-transparent, and sometimes even only translucid. Its sparkle is exceedingly brilliant, and is peculiar to itself. When found in a lesser degree in other minerals, it is designated by some mineralogists as the adamantine sparkle.

The reason of the great lustre of the diamond is, that

it reflects all the light falling on its posterior surface at an angle of incidence greater than 24° 13′. Artificial gems reflect half this light.

It is gifted with simple refraction, but to so great a degree, that Newton was induced to suspect it might be composed, in part at least, of a substance of a combustible nature.

The diamond has phosphoric and electric properties, that is, it becomes phosphoric and luminous when it has been exposed a few hours to the sun, or heated in a crucible. It is electric, inasmuch as, when heated by friction, it attracts straws, feathers, gold leaf, paper, hair—both human and that of animals, mastic, etc.; this is the case even with rough diamonds, though their surface is dull; this is a distinguishing characteristic of the diamond, which no other uncut gem possesses.

As a crystal, the diamond can only be confounded with the white spinel, which, like the diamond, is not gifted with double refraction, and is also found in the regular octahedral form, the most usual for crystals. Sometimes, however, the cube and even the dodecahedron are found, and some rare crystals unite the three. In the collection of the College of France, there are *hexatetrahedrons;* or octahedrons truncated on the angles, and consequently presenting indications of the cube. Diamond crystals are frequently hemitrope. The most beautiful form of its crystallization is the pointed octahedron.

The spinel, however, is much less hard than the diamond, and may be deeply indented by it.

Mr. Tennant relates an instance illustrative of the importance, to travellers especially, of a knowledge of the different crystallizations. For want of a knowledge of the crystalline form of the diamond, a gentleman in California had offered two hundred pounds for a small specimen of quartz. "He knew nothing of the substance, except that it was a bright shining mineral, excessively hard, not to be scratched by the file, and which would scratch glass. Presuming that these qualities belonged only to the diamond, he conceived that he was offering a fair price for the gem; but the owner declined the offer. Had he known that the diamond was never found crystallized in this form, namely, that of the six-sided prism, terminated at each end by a six-sided pyramid, he would have been able to detect the fact, that what he was offered two hundred pounds for, was really not worth more than half-a-crown*. Greater experience would have taught him that diamonds, in their natural state, are devoid of that brilliancy which is given to them by artificial means. The finest crystals of quartz are more brilliant than those of the diamond. From the inferior lustre of the latter in their natural state, it is extremely probable that numbers of diamonds are overlooked in the search for gold in Australia, California, and other gold-producing countries."

When the diamond is cut, its radiance alone suffices to distinguish it, even to the eyes of the least experienced. In the trade, however, colorless topazes, emeralds, sapphires, and the white zircon, have been passed off for diamonds. Though hardness is its chief characteristic, it is one that cannot always be tested; when it can, a fine file may be used; and if the surfaces of the stone be the least abraded or scratched by its action, the stone is not a diamond. The difference will also appear, upon close examination, without this instrument; the rays of the sun easily pass through other gems, but in the diamond they are refracted by the surface, which occasions the superior brilliancy. If the manufactured diamond is found to contain a flaw, or what is technically termed "off-color," its value is proportionately diminished, sometimes one-third,

* The same error is found in Pliny, who tells us the diamond is six-sided.

or even one-half. To ascertain if the flaw exists, the gem should be breathed on until its lustre is temporarily destroyed, its imperfections may then be easily detected.

According to Mr. Milburn, if the coat of a rough diamond be smooth and bright, with a little tincture of green in it, it is not the worse, and seldom proves bad; but if there be a mixture of yellow with the green, then beware of it—it is a soft, greasy stone, and will prove bad.

The topaz, which is powerfully electric, preserves its electricity several hours after it has been developed by friction, whereas the diamond loses this property in the course of a quarter of an hour.

Colorless corindons are easily distinguished from the diamond by their weight, which exceeds that of the diamonds in the proportion of 8 to 7. The specific weight of the diamond is nearly the same as that of the topaz. It is inferior to that of the white sapphire, the white zircon, and the white Norwegian garnet.

The diamond, when first taken out of the mine, is covered with a thick crust that scarcely permits of any transparency being visible; and even the most practised eye cannot then tell its value with certainty. It usually resembles a clear, semi-transparent pebble, well worn by the waters, a bit of unpolished glass, or of gum-arabic. Thus coated, it is called a rough diamond.

This crust is so hard that there is no substance save that of the diamond itself that can take it off. The diamond cuts every substance in nature, and can be cut but by itself.

Allusion to the dust of the diamond being the only substance that has power to lessen that gem, is touchingly made in the verses composed by Charles I., the night after his condemnation:—

> " With mine own power my majesty they wound,
> In the king's name the king himself's uncrowned;
> So doth the dust destroy the diamond!"

It seldom happens that diamonds are found in a state of natural polish, but there have been some instances. The four diamonds that adorn the clasp of the mantle of Charlemagne, so long preserved in the treasury of St. Denis, are in this primitive state.

The discovery of the art of cutting and polishing the diamond by means of its own dust, was long ascribed to Louis de Berquen, of Bruges, who is said to have constructed in 1476 a polishing wheel, which was fed with diamond dust instead of corundum, which the Chinese and Hindoos had been long accustomed to employ. Berquen was led to this discovery by observing the action produced by rubbing two rough diamonds together.

Diamonds are cut in various ways, generally with great regard to the shape of the rough stone, and assume different names in consequence; as, a brilliant, a rose, a table, and a lasque diamond. Of these the most splendid and valuable is the brilliant, from its superior sparkle, and the number of its reflections and refractions.

For the information of such as may not be skilled in the technical terms used in describing diamonds, we subjoin Mr. Jeffreys' explanation:—"The bezils are the upper sides and corners of the brilliant, lying between the edge of the table and the girdle. The collet is the small horizontal plane or face at the bottom of the brilliant. The crown is the upper work of the rose, which all centres in the point at the top, and is bounded by the horizontal ribs. The facets are small triangular spaces or planes, both in brilliants and roses. The girdle is the line which encompasses the stone parallel to the horizon. Lozenges are common both to brilliants and roses. In brilliants they are formed by the meeting of the facets on the bezils; in roses, by the meeting of the facets in the horizontal ribs of the crown. Pavilions are the under sides and corners of the brilliants, and lie between the girdle and the collet. The ribs are the lines or ridges which distinguish the several parts of the work, both of brilliants and roses. The table is the large horizontal plane or face at the top of the brilliant."

In India the chief consideration in cutting the diamond is that it lose as little as possible of its size; and rather than diminish it, they prefer leaving it of an irregular and often singular shape, the Indian lapidary having, in the number and arrangement of the facets, followed the natural shape and contours of the stone. The usual form given to them is the rose and the table.

Three hundred and fifty years ago all diamonds were cut with four flat surfaces, hence called the table or *Indian-cut* diamonds. Two hundred years later they were cut in the form of the half of a polyhedron, resting on a plane section; this was called the rose-diamond. It was not until the reign of Louis XIV. that they were cut in brilliants.

The brilliant is formed of two truncated pyramids by a common base; the upper pyramid being much more deeply truncated than the lower, the upper side of the stone presenting a table of thirty-three facets inclined under different angles, and the under side twenty-five facets. The correspondence between the upper and the lower facets is arranged so as to multiply the reflection and refraction. The first thus cut were done by order of Cardinal Mazarin, and are yet known among the French crown jewels as the twelve Mazarines.

The brilliant is the form most esteemed, as exhibiting to the best advantage the peculiar lustre of the stone; but while it ensures the best possible effect, it also entails a much larger waste of the material. Brilliants are for the most part made out of the octahedral crystals, and rose-diamonds from the spheroidal varieties.

The rose-diamond is the shape given to those stones the spread of which is too great in proportion to their depth to admit of their being brilliant cut. It is formed by covering the rounded surface of the stone with equilateral triangles, placed base to base, making the figure of a rhomb. The brilliant-mode of cutting was not much used until the middle of the eighteenth century, and the famous Jeffreys indignantly comments on "the corrupt taste that had of late (1751) prevailed, in converting rose diamonds into brilliants, under pretence of rendering them, by that means, a more beautiful and excellent jewel." Jeffreys indeed took great pains to convince the world, if not of the pre-eminence of the rose-cut over the brilliant, at least of the equality of their merits. "For," he says, "it will be found that a complete rose diamond will be more expanded than a complete brilliant of the same weight, and proportionably so in regard to spread stones . . . . And if it be admitted, as some have asserted, that there is a superior excellence in brilliants, what must be the consequence, but that rose-diamonds must sink in their value, to the great prejudice of the most noble and ancient families who are greatly possessed of them, as being a more ancient jewel than brilliants? But on the contrary, it will appear that rose-diamonds, when truly manufactured, are not inferior to brilliants, all circumstances considered."

But Jeffreys exhausted his arguments in vain; he was not more infallible on this point than in his depreciation of Brazilian diamonds, and, notwithstanding the amount of rose-diamonds possessed by the ancient families of his day, brilliants have now so completely triumphed, that they constitute the majority among the gems of the rich.

The table-cut diamond is less esteemed than the rose, and is made of those stones which, with a considerable breadth, are of very trifling depth. It is produced by a series of diminishing four-sided planes below the girdle; and the bezil is formed by one, two, or three of these planes.

Lasques are formed from flat or veiny diamonds, and are manufactured in India.

In manufacturing diamonds, that is, in converting rough diamonds into brilliants and roses, nearly half the weight is usually lost. A connoisseur will easily distinguish a modern-cut diamond from one cut thirty or forty years ago, and, all other circumstances being equal, will give the preference to the latter. Modern lapidaries seek above all things to diminish the weight of the stone as little as possible. If a hundred carats of rough diamonds yield seventy-five when cut, the profit, of course, is greater than if they are reduced to fifty. But a stone not sufficiently cut loses much of its beauty; it has neither the sparkle nor the refraction that it would have if properly cut. The setting of such stones also presents more difficulty. Some are so coarsely cut, that the most practised eye cannot distinguish the table from the pavilion; a diamond thus cut will bring less than a well-cut one of the same size, but there is more of it than there would be if better finished, and that is the chief consideration with the merchant. Formerly the utmost nicety was displayed in the cutting of the diamond; the weight was not sought at the expense of the brilliancy; all the facets were exceedingly regular, and every particle of the rough coating was carefully taken off: this is the reason why old diamonds are worth thirty to forty per cent. more than modern ones in the trade, when the quality of the stones is equal.

The art of cutting, sawing, or polishing diamonds, requires great skill, practice, and patience. "It is seldom," says Mr. Mawe, in his curious *Treatise on Diamonds and Precious Stones*, "that the same workman is a proficient in all these branches, but he generally confines himself to one. In cutting and polishing a diamond, the workman has two objects in view; first, to remove any flaws or imperfections that may exist on the stone, and secondly, to divide its surface into a number of regularly shaped polygons. The removal of flaws seems to be the most material object, since the smallest speck in some particular parts of the stone, is infinitely multiplied by reflection from the numerous polished surfaces of the gem.

When the shape of the rough stone is particularly unfavorable, the workman has to resort to the hazardous operation of splitting. When the direction in which it is to be split is decided on, it is marked by a line cut with a *sharp;* the stone is afterwards fixed with a strong cement in the proper position in a stick, and then by the application of a *splitting-knife,* the section is effected by a smart blow.

Sometimes when the section must cross the crystallized structure of the gem, recourse must be had to *sawing;* this is performed as follows:—The diamond is cemented to a small block of wood which is fixed firmly to a table, and a line is made with a sharp where the division is intended to take place; this line is filled with diamond-powder and olive-oil: the sawing is then commenced, and if the stone is large, the labor of eight or ten months is sometimes required to complete the section. The saw is made of fine brass or iron-wire attached to the two ends of a piece of cane or whalebone, the teeth being formed by the particles of diamond-powder which become imbedded in the wire as soon as it is applied to the line.

The cutting the facets on the surface of the rough stone, is a work of labor and skill; the polishing is performed in a mill.

The diamond trade is not now, as formerly, entirely monopolized by the Dutch; but the cutting and polishing of the gems is in general done in Holland, on account of the lower price of labor. The Amsterdam diamond-cutters have always been renowned. In a Jewish population of 28,000 souls, 10,000 devote themselves entirely to that trade. The General Company of Diamond Workers possesses several engines of a hundred horse-power each, setting in motion four hundred and fifty machines, and gives employment to one thousand workmen.

Though usually most esteemed and most perfect when colorless like water, the diamond is not unfrequently tinged with a slight shade of green, yellow, blue, grey, steel, and brown. There are diamonds of a perfect yellow, of a rose, and of a pistachio green hue. Some very rare ones have been found that were quite black. When they are merely tinged with a color, they are much less valuable; but if of a decided full rose, green, or other color, they are esteemed as highly as and sometimes more than the pure white. A stone of a decided and very bright color—which is very rare—if perfect also in other respects, will bring a larger price than an equally fine white one. A beautiful green stone of eight grains brought, at the sale of the Marquis de Drée, 900 francs (£36 sterling), and a white one of eleven grains, 2,000 francs (£80 sterling). Yellow and hyacinth are less valued, and blue in general least of all. A diamond of a chrysolite yellow water, of ten grains, brought 600 francs, and one of a hyacinth tint, of fifteen grains, 1,560 francs; consequently they were disposed of at much lower prices than white stones of the same weight.

One of the most beautiful specimens of colored diamonds in Europe is the blue diamond of Mr. Hope. It weighs 177 grains, and unites the most exquisite hue of the sapphire, the prismatic fire and *éclat* of the diamond. The King of Saxony possesses a beautiful green diamond, which forms the button to his state hat.

A coal black diamond was exhibited at the World's Fair of 1851, by Mr. Joseph Meyer, which was certainly a very great curiosity. It weighed 350 carats, and was so hard, that it resisted every attempt made to polish it.

Notwithstanding the advantages of the decimal-system it has not triumphed, even in France, over the carat, which is the weight used to estimate the diamond all over the world. The carat is of Indian origin; kirat is the name of a small seed that was used in India to weigh diamonds with This weight was introduced in Europe when precious stones were first brought from the East: it was universally adopted, and has always continued in use. Four grains make one carat, and six carats, or twenty-four grains, one penny-weight.

The carat grain used in weighing diamonds is different from the Troy grain, five diamond grains being only equal to four Troy grains.

As so precious an article as the diamond requires that there should be infinitely small divisions of weight, the carat is divided into sixty-four parts. This subdivision proves how difficult it would be to find a substitute for the carat.

For estimating the weight of very fine diamonds there is no fixed standard. Rough diamonds, selected as fine, and well formed for cutting, may be estimated as follows: square the weight of the stone, multiply the product by two, and the result will be the value in pounds sterling. Brilliants, if fine, may be estimated by squaring the weight in carats, and multiplying the product by eight, which will give the amount in pounds sterling.

The above rule, stated by Jeffreys and Mawe, and universally adopted by the jewellers for estimating the value of diamonds, can only hold good in the case of those that are of a small size, or do not exceed twenty carats. The value of the largest diamonds, which are exceedingly rare, must necessarily depend upon the competition of purchasers.

Diamonds are not used exclusively as articles of luxury or ornament. They are also employed advantageously in the arts. "Bad discolored diamonds," says Mr. Mawe, "are sold to break into powder, and may be said to have a more extensive sale than brilliants with all their captivating beauty. In many operations of art they are indispensable. The fine cameo and intaglio owe their perfection to the diamond, with which alone they can be engraved. The beauty of the onyx would yet remain dormant, had not the unrevealed power of the diamond been called forth to the artist's assistance. The cornelian, the agate, the cairngorm, cannot be engraved by any other substance; every crest or letter cut upon hard stone, is indebted to the diamond. This is not all; for without it, blocks of crystal could not be cut into slices for spectacles, agate for snuff boxes, etc."

All Mr. Mawe says applies to the use made of diamonds for the above-mentioned purposes at the present day; but, as to the impossibility of engraving on hard or fine stones without their aid, we have many proofs of the contrary in the fine ancient cameos and intaglios. The ancients engraved even hyaline corindons with copper tools tempered to an extraordinary degree of hardness; the secret of which the moderns neither seek nor need, as they possess steel and diamonds. We have already said that the Indian jewellers had long employed corundum for polishing the diamond; it is also a well-authenticated fact, that in 1407, that is, more than two-thirds of a century before the supposed discovery of Berquen, there lived in Paris a diamond-cutter of the name of Herman, who was celebrated in his art. The ancient Mexicans also engraved on precious and fine stones without the help of the diamond.

There are few things in the history of commerce more remarkable that the enormous mercantile value which has been set, by all civilized nations and in all ages, on gems. The jealousy with which the traders of remote times have protected this branch of commerce, and the favor with which it was regarded by semi-barbaric nations, proved the high esteem in which they were held. At the present day a very large amount of capital, in almost every part of the world, is vested in precious stones. Of this capital, however, colored gems do not appear to represent more than one-tenth, while diamonds constitute the remaining ninety per cent. With the ancients the case was reversed. Their preference for colored stones might arise in a great measure from their ignorance of the art of cutting and polishing the diamond, which reveals its prismatic hues to such advantage. The ancients, moreover, lived more in the bright light of day than the moderns, and it is by that light alone that the richness of mineral colors can be fully appreciated. The diamond, on the contrary, is far less brilliant in the broad glare of day than by night.

Nor is the diamond seen to equal advantage by every species of artificial light. The radiance of gems seen by gas or oil light shining through glass globes, is much diminished. Whoever goes to a ball where one room is lighted by gas thus enclosed, and another by candles, may notice a marvellous increase of play on the diamonds, as the fair wearers pass into the candle-lit room.

"Our present system of nocturnal illumination," says the eminent French savant, "by means of lamps, candles, gas, and even electrical light, throws over all objects hues often very unfavorable to the natural colors of gems. Thus, the sapphire, the garnet, the asteria, the turkois, the blue spinel, the amethyst, and even the opal for some of its tints, lose much by night. A striking proof of this may be had by placing a colored stone in the irradiated spectre formed by the prism with the rays of the sun. The color of the stone will then be seen to vary with the nature of the portion of the spectre by which it is illuminated in succession; and if two stones of similar colors, but of different nature, are both held in the hand, they will present a different appearance in the same kind of light. A colored strass, placed by the side of a real gem, will thus betray its worthlessness. Another experiment, more simple still, may be made: it consists in looking at a colored stone through a glass that is itself colored, red, yellow, green, or blue. Each stone will present a different appearance, and such characteristics as will indicate its nature*."

* M. Babinet: *Etudes et Lectures sur les Sciences d'Observation.*

When modern luxury is compared with that of the ancients, one of the wisest of whom did not hesitate to pay one million sesterces for a table, we shall be found inferior in all things but diamonds. Now and then a political crisis, a temporary panic, may affect the price of the gems, but their commercial value is seldom, and then but slightly and only momentarily, affected. Of all articles of trade the least subject to depreciation is the diamond. Twice within the last twenty years, however, have serious panics for a short period, greatly lowered the price of diamonds. The first was at the epoch of the discovery of the new mines of Brazil, towards the years 1843 and 1844; the second was, in France, the financial depression which was the natural consequence of the republic of 1848. The price of diamonds fell and rose with that of the government securities. That price is now at 250 francs per carat, and greatly exceeds that fixed by Jeffreys, which was £8. There is no personal property so little subject to loss, when passing from one possessor to another, as this.

"Comparing the diamond with those substances which are of the greatest utility to mankind, we find the money value of the latter to be in an inestimable ratio. Take a substance very similar to the diamond in composition, namely, coal. To express the value of an ounce of coal, we have no coin sufficiently small; it is the same with iron and lead,—metals of inestimable importance. An ounce of copper may be worth a penny; an ounce of silver may be worth five shillings; an ounce of pure gold, four pounds. But the very refuse of the diamond, that which is used for the purpose of breaking up into small particles for cutting other stones, is worth fifty pounds an ounce*."

Diamonds were once thought to be incombustible, but this has, in modern times, been proved to be an error. If exposed to a sufficient heat, the diamond will burn with an

* Professor Tennant: *Lecture on Gems and Precious Stones.*

undulating bluish flame; it will evaporate entirely in a coppel with a less degree of heat than is necessary to fuse silver, and leaves no residue.

This last property of the diamond pertains to its composition, which is pure carbon. It is not certain to whom the combustibility of the diamond first occurred. Boetius de Boot, who in 1609 published his *Treatise on Gems*, seems to have been the first who suspected this mineral might not be a stone, but a combustible body. Boyle remarked, in 1673, that when exposed to a high temperature, a portion of it evaporated. Experiments made in Tuscany and Vienna, in 1694, confirmed that of Boyle, and proved, by means of a burning lens, that fire altered the diamond by volatilizing it, and that this body did not deserve the name of Indestructible given to it by the ancients. Finally, Newton, in 1704, remarking in the diamond a power of refraction equal to that of combustible bodies, inferred that it also might be combustible.

Notwithstanding these experiments, it does not appear that the mineralogists of that day were aware of the true nature of the diamond. Macquer and Bergmann were the first to prove that the diamond was not only volatile, but also really combustible, yet they were not able to show either the cause or the result of this combustion. In 1770, very interesting experiments were made by several members of the Academy of Sciences in Paris, which resulted in the complete volatilization of the stones tried; they disappeared entirely, leaving no trace of their existence.

The instant when they become resplendent, is that when the evaporation commences; the evaporation is only on the surface, and there is no appearance of softening or fusion. If the stone is taken out during the operation, the portion that remains will be found to have lost none of its primitive properties; its size and weight alone are diminished.

The effects of this combustion, first examined by Lavoisier, in 1772, were subsequently by Guyton de Morveau, in

1785. The latter gentleman exposed a diamond to intense heat, enclosed in a small cavity in a piece of iron. When the cavity was opened, the diamond was entirely gone, and the iron around was converted into steel. This proved it to be *pure* carbon, which combines with iron to form steel, and not *charcoal*, which is generally an oxide of carbon. The peculiar hardness of steel is to be ascribed to its union with a portion of pure carbon, or diamond.

It is no uncommon thing for jewellers to expose such diamonds as are foul to a strong heat, embedded in charcoal, to exclude the atmospheric air, otherwise the intense heat would produce combustion.

In reality, carbon is indestructible: man may alter its *form*, converting the brilliant adamant into an invisible gas, but still not a particle is lost; and if charcoal were heated for centuries in a vessel from which the air were excluded, no change would be perceptible.

That charcoal and the diamond are similar substances in their chemical nature, differing only in mechanical texture, has been proved by an experiment of Sir Humphrey Davy. He exposed charcoal to intense ignition, *in vacuo*, and in condensed azote, by means of Mr. Children's battery, when it volatilized and gave out a little hydrogen. The remaining portion was always much harder than before, and in one case so hard that it scratched glass, while its lustre was increased; but it remained black and opaque.

French chemists have made this experiment with equal success. That of the combustion of the diamond was renewed recently by M. Dumas, with a view to establish with certainty the atomic weight of the carbon. He again proved that the diamond, being exclusively formed of carbon, the almost imperceptible residue it leaves after its combustion, is the result of foreign minerals disseminated in its mass.

Mr. Cross also made an interesting experiment. Having obtained water from a finely crystallized cave at Holway, he succeeded in producing from that water, in the course of ten days, numerous rhomboidal crystals, resembling those of the cave. In order to ascertain if light had any influence in the process, he tried it again in a dark cellar, and produced similar crystals in six days, with one-fourth the voltaic power. He repeated the experiment a hundred times, and always with the same results. He was fully convinced that it was possible to make even diamonds; and that at no distant period, every kind of mineral would be formed by the ingenuity of man. By a variation of his experiments he had obtained grey and blue carbonate of copper, phosphate of soda, and twenty or thirty other specimens. A French savant, M. Despretz, has, by means of the arc of induction, and by weak galvanic currents, after three months of slow and continuous action, produced crystallized carbon in *black octahedres*, in *translucid colorless octahedres*, in *colorless* and *translucid lamina*, possessing *the hardness of diamond dust*, and that disappeared by combustion, *leaving no residue*. M. Gaudin, having examined the result of M. Despretz's experiments, pronounced them to be *microscopic diamonds*.

Science has made extraordinary progress within the last century; with regard, however, to the formation of diamonds by human ingenuity, that is yet far beyond her reach.

During the age of Louis XIV., the possibility of making diamonds grow larger by depositing them in certain liquids, just as crystals of salt are made to increase in size in a solution of the same kind, was implicitly believed in. Every one knows the story of the gold pieces manufactured by M. Sage with substances extracted from the ashes of burnt vegetables. The result was splendid as a scientific one, but not a very lucrative one, since each twenty-franc piece cost a hundred and twenty-five francs in the extraction.

There is an immense difference between the analysis of

bodies and their synthesis. While the generality of bodies may be submitted to analytical processes, there are numerous inorganic bodies that resist all the chemist's efforts in the converse operation; and with regard to the different forms of organized matter, he finds himself incapable of bringing back that of which he knows the exact component parts, and which he found no difficulty in decomposing. He may reduce a fine diamond to a small amount of carbon, but of that carbon he cannot reconstruct the gem he has destroyed. No, Nature keeps her secret well; she is the only diamond manufacturer, and she carries on her operations in a laboratory to which man has no access, and his ambition and pride must recoil before her insurmountable barriers.

Thus, though it is not likely we shall ever see a diamond of a carat the work of man, there remains one fact well proved, namely, that a small quantity of the gas called *carbon*, which, when in an aëriform state, destroys life, produces, when acted upon in different ways in the great laboratory of Nature, two substances perfectly unlike each other,—charcoal and the diamond,—the one consumed as fuel, the other prized at so high a rate as not only to be purchased for sums equal to princely fortunes, but sought at the cost of the blood of many thousand human beings, the total ruin of empires, and the sacrifice of another far more costly jewel—honor.

CHAPTER II.

CHEMISTRY OF PRECIOUS STONES.

Crystal.—Former Belief regarding its Composition.—Classification of Gems.—How to distinguish White Hyalines from Diamonds.—Composition of Corindons.—The Ruby.—The Sapphire.—The Amethyst.—The Topaz.—The Beryl.—The Emerald.—The Chrysolite.—The Peridot.—The Cymophane.—The Garnet.—The Hyacinth.—The Opal.

" Des métaux récents dont l'art fit la conquête,
Chacun a son pouvoir: le chrome est à leur tête.
Peintre de nos minéraux, de nos plus belles fleurs
Il distribue entre eux les brillantes couleurs;
L'émeraude par lui d'un beau vert se colore;
Il transmet au rubis la pourpre de l'aurore;
Quelquefois du plomb vil fidèle associé
Teint d'un vif incarnat son obscur allié."

DELILLE: *Les Trois Regnes de la Nature.*

CRYSTAL was in all probability the first substance ever noticed as occurring in a regular form. From its extreme transparency, the ancients believed it to be water permanently congealed by extreme cold, and hence called it *krustallos*, signifying ice. The word has since been applied to all the regular forms of minerals without reference to its original meaning.

The crystals used by *bijoutiers* and jewellers are more or less rare, more or less precious, and consequently more or less dear. Some are merely quartzose, sometimes opaque and sometimes transparent and white; or, presenting the most vivid and varied hues, are exclusively composed of silex, and are of little value; others are finer, have a peculiar lustre, will scratch glass, and are more valuable in proportion as they are more rare, more difficult to work, and that their hues are vivid, and their sparkle great.

The first and less valuable are employed by the *bijoutier*, and wrought as low-priced trinkets. The second are usually mounted by the jeweller in rings, bracelets, necklaces, and other ornaments, which frequently represent princely fortunes.

The stones usually worn may be classed as follows:

Diamonds; hyaline corindons and other precious stones; hyaline quartz, or rock crystal; feldspath, jade, lapis-lazuli, malachite, and fluorine: siliceous hydrated stones, possessing simple translucidity, such as agates; siliceous hydrated stones, entirely opaque.

The precious stones that rank immediately after the diamond, are termed by mineralogists hyaline corindons, or telesia. They are almost always diaphanous, though sometimes only translucid. Almost exclusively composed of alumina in the proportion of from 90 to 98, their opacity increases (or their transparency diminishes) according to the proportion of iron or silex they contain, to which latter elements they owe their tones.

The hyaline corindons are either white or colored, and are named accordingly.

The crimson red	corindon	is the	oriental ruby.
The colorless	„	„	white sapphire.
The azure blue	„	„	oriental „
The indigo blue	„	„	indigo „
The violet	„	„	oriental amethyst.
The yellow	„	„	„ topaz.

When the corindon is perfectly colorless, it is extremely brilliant, and has sometimes passed for a diamond. The hardness of the latter mineral, superior to that of the corindon, suffices to distinguish it. However, as we do not like to scratch a gem, and as to repolish it would diminish its thickness, the specific weight is usually made the criterion. This is not only a test for the diamond, but also a sufficient one to distinguish all colored corindons from the stones which usurp their names. There is, however, a safer and easier method of distinguishing the white sapphire, or the topaz, sometimes offered for sale as a diamond. This is done by observing if the stone possesses the double refraction, in which case, it is perfectly evident the gem is not a diamond. This optical character is explained by M. Babinet, as follows:—"If, when we look through a stone at any minute article, such as the point of a needle, or a small hole in a card, we see the article doubled, as though two needles or two holes were looked at instead of one, the stone refracts a double image. All white gems possess this property except the diamond. A little practice and dexterity being required in order to make this examination, the stone and the mineral should be fixed by means of a bit of wax to some slight support, that both may be steadied. This simple experiment may be tried by any one, and does not require that the stone be taken out of its setting." The white Brazilian topaz so greatly resembles the diamond, that it is only by testing its properties that the difference can be ascertained. This external similitude gives to the colorless sapphire and topaz a higher price than they would otherwise have.

The hyaline corindons are, after the diamond, the hardest bodies in nature; they can scratch any other stone, but can be scratched but by one of their own class.

Strange as the assertion may seem, it is nevertheless true, that nature has formed the most precious bodies of the most common materials; thus the clay called by chemists *alumina*, and the rohite pebble, or rock-crystal, called *silex*, form the basis of all precious stones. The opal is formed of pebbles and water; the topaz unites a small portion of fluoric acid to silex and alumina; the emerald, the chrysolite, the beryl, the tourmalin, contain also another element, *glucina*; and the garnet is so ferruginous that it acts on the needle.

In their commercial value, the hyaline corindons differ in one respect from the diamond, namely, that the latter, from the smallest specimen to the regal gems of world-wide celebrity, has, like the precious metals, a price proportionated to its weight. Very small hyaline corindons, on the contrary, are of little value; they only begin to be appreciated when their size, rendering them rare, ensures a high price. Thus, in order that the pivots of watches may turn with greater precision, they are set in small rubies perforated for the purpose. These minute gems, about the size of millet grains, though very useful, are very low priced on account of their abundance; but, when a perfect ruby of five carats finds its way into the trade, it will bring twice the price of a diamond of the same size; and for a ruby of ten carats, three times the price of a diamond of the same weight will be offered for it, though the diamond be valued at eight hundred or a thousand pounds.

THE RUBY.

The very name of the ruby, the most beautiful of hyaline corindons, called by the Greeks anthrax, "live coal," betokens its color, a vivid blood-red. A perfect and large sized ruby is universally acknowledged to be one of the rarest productions of nature. The ruby loses none of its exquisite beauty when viewed by artificial light. The brilliant color of the ruby is attributable to a sixth part of it being chromic acid, while other gems sometimes mistaken for it, such as the garnet, are colored by oxide of iron. The most esteemed, and at the same same time the rarest, color, is pure carmine, or blood-red of considerable intensity, emitting, when well polished, a blaze of the most exquisite and unrivalled splendour. It is, however, found in paler hues, and sometimes mixed with blue in various proportions; hence there are rose-red, (a beautiful variety of the oriental ruby,) reddish, white, crimson, peach-blossom, red, and lilac-blue. The latter is called the oriental amethyst.

There are four different kinds of rubies; the cochineal-red, or *oriental* ruby, the *spinel* ruby, the *balass* ruby, and the *rubicelle.*

The oriental ruby, under the name of carbuncle, plays a great part in Eastern legends, and in ancient romances; its blood-red hues illumined enchanted halls, and led valiant knights to the rescue of dragon-guarded damsels. It had the marvellous property of shining in the absence of all light. The real truth is, the ruby may be rendered phosphorical, like all other precious stones, by exposing it to the rays of the sun, or by putting it in a close crucible, heated to a certain degree.

The ruby figures under the name of carbuncle among the twelve gems that adorn the breast-plate of the High Priest; yet, as there is no evidence that the ancients engraved on the ruby, the stone mentioned is by many supposed to have been the oriental garnet, a transparent red gem with a violet shade and strong vitreous lustre. Upon the whole, there is great uncertainty as to what really were the gems denoted by the Hebrew words that are understood to signify the ruby, the emerald, etc.

The second class ruby is the spinel. This variety, which must not be confounded with the corindon or telesia of mineralogists, is, when perfect, scarcely less beautiful than the oriental, though less rich in its color, which is a full carmine, or rose-red.

The third class ruby, the balass, is of a pale rose, or sometimes full rose tint; the last is the most esteemed. The commonest color is a pale red. They are only esteemed perfect when of a certain weight.

The Brazilian topaz is often made to imitate the ruby, by placing it in a crucible filled with sand and heated to a certain degree. The topaz is thus made to lose its orange-yellow tint and acquire the exquisite rose hue of the balass.

The fourth class ruby, *the rubicelle*, is an inferior variety of the spinel; its hue is a yellowish red.

The finest crystallization of the ruby is not, like that of the diamond, the octahedron; it is the six-sided prism like rock-crystal, but without the six-sided pyramids that terminate the prism of rock-crystal. It is easily frangible, and is infusible before the blow-pipe. Its specific gravity is the same as that of the sapphire, 4 to 4 2.

The jealous care with which the East Indian princes have prevented fine rubies from being carried out of their domains, makes those of a certain size very scarce. If perfect, a ruby of three or four carats sells for more than a perfect diamond of the same weight. When the cabinet of precious stones of the Marquis of Drée was sold, a very fine diamond of eight grains (two carats) brought £32; while for a ruby of exactly the same weight £40 was paid. At the same sale the price of a ruby of ten grains rose to £56.

THE SAPPHIRE.

This gem is so closely allied to the ruby, that many mineralogists pronounce them varieties of the same stone, the blue being only rather harder than the red. In its purest state, the sapphire is of a clear beautiful azure colour; unfortunately, it loses much of its delicate beauty, seen by artificial light. The hue of the perfect sapphire is soft, rich, velvety, neither too light nor too dark. Such was the sapphire the ancients consecrated to Apollo.

The clear bright indigo-blue sapphire is also much esteemed.

This hyaline loses value according as it is more or less clouded. Sometimes also it is only transparent in parts, being spotted and streaked with a deeper color in others. Several magnificent sapphires, from the collection of the late Mr. H. P. Hope, were seen in the London Exhibition. Mr. A. J. Hope exhibited his well-known "*Saphir merveilleux*," which is blue by day, and amethystine by night.

This gem afforded the foundation for one of Madame de Genlis' stories.

The asteria, or star-stone, is another variety of the sapphire, which presents the form of a star, with six radii, all of which sparkle with great vividness, as its position is varied in the sun. It is semi-transparent.

The sapphire is infusible before the blow-pipe. It becomes electrical by friction. A sapphire of ten carats weight is considered worth fifty guineas, and one of twenty carats, two hundred pounds.

THE AMETHYST.

The *oriental* amethyst is a gem of the most beautiful violet, and the rarest of all the hyalines; so rare is it, that before the Revolution of 1789, Mr. d'Augny, the wealthy financier, was the only private person known to possess a perfect and beautiful amethyst. It is as hard as the ruby and sapphire, with which it also corresponds in form and specific gravity. It is of so remarkable a richness of color, and takes so brilliant a polish, that many mineralogists are tempted to call it a violet sapphire, and place it in the class of colored diamonds, or rubies.

The *occidental* amethyst is merely crystal or quartz, colored with peroxide of iron, or manganese. It is frequently defective in color, being tinged only at one extremity, while the other is quite white. When perfect, however, it is very beautiful, and of a rich hue, like that of the purple grape. It will take a brilliant polish, and greatly resembles the oriental gem. It is found in rolled pieces, in the alluvial soil, in the interior of balls, or geodes of agates, and finely crystallized in veins and cavities of grünstein and other rocks.

The price of the oriental amethyst can scarcely be specified, on account of its extreme rarity; and the western are sold at a valuation, in accordance with their size, beauty, and richness of color. The amethyst is the stone set in the rings of bishops. The western occurs of considerable size, and the ancients used it, not only for personal ornaments, but had cups of amethyst, which they prized highly.

THE TOPAZ.

The topaz is a gem of a truncated, prismatic form, the hardest of all hyaline corindons, with the exception of the ruby and sapphire. Its specific gravity is about 4.

The Emperor Maximilian, who, with a blow of his fist could knock a horse's teeth out, and with one of his imperial kicks, break the animal's thigh, is said to have crushed topazes to powder between his fingers. Whatever was the gem to which the Romans gave the name, it certainly could not have been the hyaline corindon, so called now, even making all due allowance for the Thracian Goliath's strength.

This exquisitely beautiful gem is of a bright *jonquil* or citron, at times of a bright, clear gold color. The most esteemed is the soft satin-like hue, which seems full of resplendent gold spangles. The color should be perfectly clear and equal, but the tint is different, according to the place whence it is brought. In the trade, the Egyptian topaz is sometimes passed off for the oriental, but it is less hard.

The Brazilian topaz is, after the oriental, the most esteemed, and the hardest; it is of a deep orange tint, and will take the finest polish.

The Bohemian topaz is of a deep hyacinth, and sometimes a brown tint.

The Saxon topaz is very hard, extremely brilliant, and of a clear, transparent yellow. It loses its color when subjected to the heat of a small crucible, or when placed in a tobacco-pipe, and covered with ashes or sand and heated; it then becomes perfectly white.

Topazes are also found in the Voigtland, two miles from Auerback, in the quartz, or among the crystallized free-stone, and sometimes covered with a yellowish marl.

There are also other stones called German topazes, but they are only a sort of glassy spar that can deceive no one. It is erroneously supposed that these stones take their yellowish tint from lead, the more so as lead is used to color crystals to imitate topazes.

The price of colored gems varies with the caprice of fashion, and it is difficult to determine what that may be, with regard to these beautiful ornaments, for any length of time. For diamonds the price varies little.

The topaz is not always yellow; there is the *red*, which is a very beautiful variety, sometimes mistaken for the ruby. The *blue* topaz is also a beautiful gem, of a fine azure blue tint.

THE BERYL.

This gem is found in stones of larger size sometimes than any other of the corindons. Stones of one, two, or even three ounces are not uncommon, and a gigantic opaque beryl from North America, weighing eighty pounds, was in the London Exhibition; there is another of about the same size in the British Museum.

There are two kinds of beryls; the oriental and the western.

The crystallization of the beryl is polygon; it is transparent, of a bluish green, and sometimes of a pale sea-green. This stone was suspected by Pliny to be a species of emerald, a conjecture which modern mineralogists have completely confirmed; both have the same crystalline form, hardness, and specific gravity. The term "emerald" is applied to that particular variety which presents its own peculiar color, or *emerald* green; while that of beryl is given indiscriminately to all the other varieties; as the sea-green, pale blue, golden yellow, and white.

The *oriental beryl* is beautifully transparent and exceedingly brilliant; its color is delicate, partaking of blue and green mingled in equal proportions. The emerald is perfectly green without mixture of blue. The sapphire is a pure blue without a shade of green, whereas the beryl participates of both colors with an infinite number of shades. The oriental beryls are, however, more fixed and decided in their shades, harder and more susceptible of a fine polish, than the western. These last also present a very pretty mixture of blue and sea-green, and will take a fine polish. Artificial light, that deprives the beautiful sapphire of its ethereal hue, adds greatly to the brilliancy of the beryl.

THE EMERALD.

The beautiful color of the emerald is due to one or two per cent. of oxide of chromium. This gem consists of silica sixty-eight, alumina fifteen, glucina, with a trace of lime, oxide of iron, and chromium. The emerald breaks readily at right angles to its *axis*. Its crystallization is a hexahedral prism truncated at both ends.

The emerald, to be perfect, should be of a rich, soft, lively, pure, meadow-green, and without blemish; but this gem is seldom found without a flaw when of some size. They are usually obscured by cloudy spots that often and sometimes totally destroy their reflection. An emerald without a flaw has passed into a proverb. Fine emeralds are so rare, and in such demand, that, according to Mr. Mawe, a particular suit has been known to have passed into the possession of a series of purchasers, and to have made the tour of Europe in the course of half a century.

This gem is the lightest of all hyalines, its specific gravity being only from 2=6 to 2=7·7, like rock-crystal; while that of the diamond and topaz are 3½, and that of gems of the sapphire family, 4, and the zircon and garnet still higher. It is so soft, that it hardly scratches crystal. It is found in crystals of a beautiful green, embedded in a species of whitish clay. The presence of these deposits in a stone of a nature and color so foreign to the emerald, can only be accounted for by admitting electricity as the cause*.

The emerald has always been a favorite with the public, and takes, as such, a place immediately after the ruby. It has moreover this advantage, in common with the ruby, that it does not lose its beauty by artificial light. It is usually surrounded by brilliants, and thus set, forms one of the most charming ornaments that can be worn.

The finest emerald known was at the Great Exhibition of the year 1851. It is two inches in length, and measures across the three diameters two and a quarter inches, two and one-fifth of an inch, and one and seven-eights of an inch, and weighs eight ounces, eighteen pennyweights.

Pebbles of quartz are frequently mistaken for beryls, and *vice-versá*. When these substances are crystallized, the means of distinguishing them are very simple; quartz is striated longitudinally; and by sacrificing one or two crystals, and observing the fracture, the truth may be ascertained; if emerald or beryl, the fracture will be in planes, or the cleavage of topaz; if quartz, the fracture will never be in a straight line, but conchoidal †.

THE ORIENTAL CHRYSOLITE, OR CHRYSOBERYL.

The oriental chrysolite is a gem that crystallizes in an oblong hexahedral prism with unequal sides, ending in two tetrahedral pyramids. It is of a beautiful clear apple-green, and will take a very fine polish.

There are two sorts of Brazilian chrysolites. One something like the oriental peridot in color, save that it remains

* M. Babinet. † Professor Tennant.

tinged with yellow, and that it is somewhat darker. It is less hard than the peridot, and will not take so fine a polish.

The other Brazilian chrysolite is straw-colored, tinged with a fine shade of green. This chrysolite is very hard, and will take a very fine polish. Very different mineral substances have been called by this name; the chrysoberyl, the peridot, certain varieties of beryls and topazes, have borne the name of chrysolite. Its fracture is conchoidal. It is composed of 80·25 alumina, and of 19·75 glucina. The largest specimen of this stone is in Rio Janeiro. It is found in Ceylon, Pegu, and Brazil.

THE PERIDOT, OR OLIVINE.

The peridot is a very pretty olive-green stone; it is so soft that it will scarcely scratch glass. It is often found, but in very minute crystals, in the lava of volcanos, and, moreover, has the singular honor of being the only gem that has hitherto been found among the stones that have fallen from the moon, or rather, the atmosphere. The fine oriental peridots that are brought from India, are very beautiful ornaments, and it is to be regretted that they meet with so little favor. It is composed of silicate of magnesia, (silicate 43·7, magnesia 56·3,) with a certain quantity of protoxyd of iron. The crystallization is rhomboidal.

THE CYMOPHANE, CAT'S-EYE, (BELL'OCCHIO.)

This stone, which is brought from Ceylon, is of a greenish hue, striated with filets of white amianthus. It sometimes possesses a milky and opalescent appearance. When cut, *en cabochon*, it exhibits a floating white band of light, produced by the reflection of the light on the fillets of amianthus. Some varieties are called cat's-eye, because, when exposed in a certain position to the light, they emit one or more brilliant rays, colored or colorless, interiorly or on the surface, issuing from one point, as from a centre, and extending to the edges, and disappearing when viewed in another light. Many precious stones have at times this quality.

THE GARNET.

There is no gem that varies so much as the garnet, whether in depth and diversity of color, or in variety of shape and crystallization. There are dark-red, yellowish, dark-brown, orange-yellow, and even black.

Garnets, in general, have neither the brilliancy nor the transparency of other gems. From this rule must be excepted the oriental or Syrian, so called because brought from Syrian, or Sorian, the capital of Pegu, which is probably the oriental, the color of which is a rich blood-red. This garnet, like the spinel ruby, with which it is sometimes confounded even now, was called by the ancients, *carbunculus*. The Syrian garnet is sometimes of a fine violet color, which, in some rare specimens, makes it compete with the amethyst; from the latter gem it may, however, be discriminated by the disadvantage of losing its brilliancy, and acquiring an orange tint by candlelight. Distinct from all other garnets, it preserves its color, unmixed with the common black tinge, unassisted by foil, even when thick.

Garnets are composed of silex, alumina, and a small proportion of oxide of iron, from which they derive their beautiful color. When there is an excess of it, their play and splendour is injured, and the color is of a reddish brown. The perfection of all gems depends less on the quality of their component parts, than on their complete solution and intimate combination. The alkalized earths, as lime, magnesia, and, still better, potash, seem to intervene as solvents, for alumina completely dissolved, acquires a crystallization of which, by itself, it is not susceptible.

Garnets differ greatly in size and hardness. Some are like a grain of sand, others the size of an apple; some will scratch quartz, others, on the contrary, can be scratched by quartz; some are transparent, others opaque. The primitive form of the garnet is the rhomboidal dodecahedron. When garnets are of a fine color, transparent, and capable of receiving a fine polish, they constitute a very pretty ornament, and are cut *en cabochon*, or in facets. The garnet is infusible in the blow-pipe and melts down to a black enamel.

The so-called *western* garnet, which is merely an inferior gem, is of a deep red, more or less bright, according to its hardness. It differs from the red oriental, inasmuch as its color is less soft, and is not tinged with velvet or purple; it also loses its color by candlelight and looks black.

The *Bohemian* garnet is, however, of a fine bright blood-red, and sufficiently hard to take a fine polish. There are some of great size and brilliancy. The refraction of the garnet is simple, like that of the diamond. It is cut *en cabochon*, and will sometimes have the effect of an asteria, or a four or six-rayed star, like the generality of the corindons.

THE HYACINTH,

is a gem of the family of the garnets, though Haüy has thought proper to separate it as of a different nature. To be perfect, it should be of a beautiful orange with a shade of scarlet, quite clear, and without flaw; it is then denominated the *beautiful* hyacinth. It was a great favorite with the Romans, but is little worn at present. The western hyacinth comes from Brazil in quadrilateral crystals, terminating at both ends in a pyramid, with the same number of facets.

The real and rich oriental hyacinth is the zircon, of a deep honey tint. If smoked rock-crystal be heated, it takes a rich hyacinthine hue. The essonite of Haüy, which is a species of garnet, is also a hyacinth.

THE OPAL, (HYDRATED QUARTZ.)

This beautiful gem, so justly prized by the Romans, is also a favorite with the moderns. Fashion at the present day ranks it before the sapphire, and, when surrounded by brilliants, as it is usually set, the bright scintillating rays of the diamonds contrasting with the calm and soft, but rich and varied tints, of the opal, it is one of the most beautiful ornaments imaginable. No better and more precise description of this stone can be given than that of Mr. Mawe: "The color of the opal is white or pearl-grey, and when seen between the eye and the light, is pale red or wine yellow with a milky translucency. By reflected light it exhibits, as its position is varied, elegant and iridescent colors, particularly emerald-green, golden-yellow, flame and fire-red, violet, purple and celestial blue, so beautifully blended and so fascinating as to captivate the admirer. When the color is arranged in small spangles, it takes the name of the harlequin opal. Sometimes it exhibits only one of the above colors, and of these the most esteemed are, the emerald-green and the orange-yellow."

The precious opal is not quite so hard as rock-crystal; it is frequently full of flaws, which greatly contribute to its beauty, as the vivid iridescent colors which it displays, are occasioned by the reflection and refraction of the light which is decomposed at these fissures. It is never cut in facets, but always hemispherical. It is generally small, rarely so large as an almond, or hazel-nut, though I have seen some specimens the size of a small walnut, for which several hundred pounds were demanded.

Opals which have the quality of emitting various rays with particular effulgence, are distinguished by lapidaries by the epithet *oriental*, and often by mineralogists by the epithet *nobiles*. This opal, also denominated the spangled, in which the rays of the sun are reflected like so many colored brilliants, is held in high esteem by the East Indian princes, who consider it the equal of the diamond.

The opal which the Roman senator carried into exile with him, was estimated *sestertium* viginti millibus (3,881,000 francs, or £155,240). It was about the size of a hazel-nut. As the Romans gave to the opal the rank the diamond occupies with us, this price might not be considered too high for a gem that was probably the *Regent*, or the *Koh-i-noor*, of Rome. Before the Revolution of the last century, the financier d'Anguy possessed a harlequin opal of great beauty. This stone, perfect in every respect, and measuring twenty-one millimetres in length and fifteen to sixteen in breadth, was celebrated; but its possession did not cost its master as heavy a penalty as that of Nonius.

There are black opals that come from Egypt, but these are very rare. They have the glow of the ruby seen through a vapour, like a coal ignited at one end.

When first taken out of the earth, the opal is very soft, but it hardens and diminishes in bulk by exposure to the air. The greatest care is required in the cutting, or it is easily spoiled, as it is one of the softest of gems, though its hardness varies considerably. Its specific gravity is 19·58 to 2·54. The lowness of its specific gravity, in some cases, is to be ascribed to accidental cavities which the stone contains, and which are sometimes filled with drops of water Its colors are not occasioned by any particular tinge of substance, but by the peculiar property these small fissures have of refracting the solar rays. It is a compound of 90 silica, and 10 water. There is an opal called *hydrophane*, which is white and opaque till immersed in water; it then resembles the former.

CHAPTER III.

CHEMISTRY OF GEMS.

Hyaline Quartz, or Rock-crystal; its composition.—The Turkois.—Agates.—The Onyx, Sardonyx, Chalcedonyx, &c.—Jasper.—Feldspath.—Lapis-lazuli.—Malachite.—Amber.

QUARTZ is exclusively formed of silica, with a very slight portion of alumina. Hyaline quartz, or rock-crystal, is quite colorless. Though heavier and much harder than artificial crystal, it bears a perfect resemblance to it. It is found in crystallized rocks, and in lodes; it is sometimes washed up from primitive deposits in the beds of rivers: the purest and clearest comes from Madagascar; very fine is also found in the Alps. Rock-crystal, which was formerly much prized, has been almost entirely superseded by artificial crystal. From the difficulty of cutting it, the articles made of it, such as chandeliers, vases, seals, and buttons, were very high-priced; the same things are now made much easier, and consequently at less expense, of artificial crystal. Fine crystals of quartz are still used as substitutes for glass in spectacles, and various optical instruments, on account of their superior hardness, which renders them less susceptible of being scratched.

The tinges quartz receives from metallic oxides produce all the various colors we admire in the hyaline corindons, or precious stones. The false amethyst, or purple quartz, is tinged with a little iron, or manganese. *Rose-quartz*, or the false ruby, derives its color from manganese. Aventurine is a beautiful variety of quartz of a rich brown color, which, from a peculiarity of texture, appears filled with bright spangles. Small crystals of quartz tinged with iron are found in Spain, and have been termed "Hyacinths of Compostella."

Chalcedonyxes, cornelians, onyxes, sardonyxes, blood-stones, or heliotropes, and the numerous varieties of agates, are principally composed of quartz, tinged with various other substances.

Very beautiful brown and yellow crystals of quartz are found in the mountain of Cairn-Gorm, in Scotland. They are made use of for seal-stones and other trinkets, and sometimes improperly called "topazes."

Crystals tinged with all the colors of precious stones are also found in Alençon, Medoc, and on the borders of the Rhine.

Whatever color the hyaline quartz may be tinged with, it cannot compete with artificial crystals, which are also made of any color required. Thus the colored hyaline quartz is in much less demand than the white.

Among the semi-opaque and opaque stones, first in rank, in beauty, and in value, is

THE TURKOIS.

So highly esteemed has this gem always been among ancients and moderns, that it takes a place among precious stones.

There are two kinds of turkoises bearing some resemblance to each other, but differing entirely in nature. The oriental, or *Vieille Roche* turkois, which is the real gem, the Calaïs of the Romans, offers very marked exterior characteristics; it is of a beautiful azure-blue or bluish green, opaque, or slightly transparent on the edges; its hardness, which is rather greater than that of phosphated lime, permits of its receiving some polish; its specific weight varies from 28, 36 to 30; it is infusible in the blow-pipe and cannot be attacked by acids.

The *ensemble* of the characteristics of the turkois plainly distinguish it from all other blue mineral substances found in nature; but its chemical nature has hitherto remained problematic, and the different analyses hitherto given have produced results very much at variance with each other. The only elements presented invariably were phosphate of alumina, oxide of copper, and iron.

This gem is much esteemed, and always brings a high price if the color is fine. An oval turkois, five lines by five-and-a-half, light blue, with a slight greenish tinge, has brought twenty pounds at a public auction. Still this gem must have greatly diminished in price since Shakspeare's time; or Shylock perhaps valued it more for imaginary virtues, or as a memorial of his wife, than for its pecuniary value, when he expresses such deep regret at its loss.

The high price of the turkois has occasioned the use of various substances colored blue, and presenting the same appearance when mounted as ornaments. The most commonly used are the teeth of fossil mammalia colored, it is said, with phosphate of iron, found in Auch, department of Gers, France. The hardness of these mammiferous teeth is much less than that of the turkois; when touched by acids, they effervesce, and when burned, diffuse an animal odour, all which characteristics distinguish the two substances easily, even when the color leaves a doubt.

AGATES.

Agates are a variety of hydrated quartz, composed of a fine, compact, unctuous, and translucid paste, which will take the finest polish. They are never wholly opaque like jasper, or transparent as quartz crystal; they take a very high polish, and the opaque parts usually present the appearance of dots, eyes, veins, zones, bands, etc. The colors are usually reddish, bluish, yellowish, milk-white, honey-orange or yellow, flesh, blood, or brick-red, dark brown, violet, blue, and brownish green. They are found in irregular rounded nodules, from the size of a pin's head to more than a foot in diameter.

The oriental agate—so called from its superior beauty —is semi-transparent. If the color is a milky-white, or tinged with yellow or blue, it is a chalcedonyx; if tinged with orange, a sardonyx; if rose-colored or red, a cornelian; if a beautiful clear green, a chrysoprase. There are agates the ground of which is grey, with fillets twisted into spirals.

The *leonine*, or *panther*-agate, is brown, with spots like a panther, or waves of a deeper shade.

There are agates with red veins, others with white ones. There are also fine black agates. On some agates may be seen figures of shrubs, birds, mosses, clouds, stars, animals.

THE ONYX.

The name of this gem is from a Greek word signifying *nail.* The onyx is a semi-transparent agate, exhibiting layers of two or more colors strongly contrasted. A sard, or sardoin, having a layer of white upon it, is considered an onyx. The price of the onyx increases according to the number of zones or layers, to the distinctness of the colors, and to the marked contrast they present. The most famous Greek engravers selected different varieties of onyxes for their masterpieces, on account of their hardness, which renders them susceptible of receiving he finest polish.

In the three-banded onyx, the upper layer is generally brown, and when held between the light and the eye, presents a reddish tint. The second layer is of a white or milky grey; the third, a beautiful black or smoke-color. The zones should be very distinct, separate, and strongly marked. The more lively the colors, the more rare and curious the stone. Sometimes there are but two colors;

the upper, a beautiful white, the inferior grey; the stone is then called an agate onyx. The most valuable are those with four distinct bands.

The *sardonyx*, or *sardius*, is an onyx, one of the zones of which is red, and the other some color peculiar to the onyx. The word *sàrd* designates the red color of the cornelian.

The oriental sardonyx comes from India, Egypt, Arabia, Armenia, and Babylonia. The western, called by the Italians *niccolo*, is covered with bluish spots, and surrounded by milky zones. It is less hard than the oriental, and is chiefly brought from Bohemia.

There is a sardonyx presenting several zones, in the common centre of which there is a spot somewhat resembling an eye. The artist takes advantage of this accident to enhance the effect; the stone is perforated underneath the spot, and a bit of gold foil is introduced in the cavity, giving great *éclat* and play to the spot.

The *cornelian* designated by Pliny is a sort of semi-transparent agate, of a fine grain, and susceptible of a very high polish. It is usually red, or blood-colored, sometimes flesh-color. The oriental cornelian is a table ruby, of a fine brownish red, and very hard like the other corindons. The enormous Indian rubies are often only masses of red corindon without transparency. It comes from the Indies, Arabia, Persia, and Egypt. There is a white cornelian with a bluish or milky tint; a red or yellow, with white, red, or black lines. Others again are spotted as with gouts of blood.

The *chalcedonyx* is a semi-transparent, cloudy, milk-blue stone. The sapphirine chalcedonyx is the hardest, handsomest, and most esteemed; it is of a bluish grey. There are others in which there is a shade of blue or purple; these three colors, when the stone is in the sun, multiply and present all those of the rainbow, whence it is called the *iris chalcedonia*. The most common is milk-white, and only distinguished from the white agate by its being more clouded and less hard. There is the striped, the spotted, etc. etc. The name of chalcedonyx is given to all fine stones of the agate kind, the transparency of which is lessened by a cloudy shade.

JASPER.

The different varieties of jasper are distinguished from the agates by their lustreless fracture and complete lack of transparency; they are used to great advantage in mosaics on account of the variety of their colors.

FELDSPATH.

Feldspath is a very pretty substance, used for a variety of fancy articles. It is characterized by the opalescent or milky tints that emanate from its interior, and by its color, which varies according to its transparency and reflection. It is composed of silex, alumina, and potash, and is found abundantly disseminated in almost every country. White and less hard than quartz, it is fusible in the blow-pipe. It is found of various colors, sometimes opaque and sometimes semi-translucid. Some varieties are truly remarkable; and moon-stones, sun-stones, amazons, and labradorites, when very fine, bring high prices. The moon-stone, is translucid, and resembles a milk-white opal faintly reflecting the light. The sun-stone is of a yellowish color, and covered with an infinity of bright gold dots, which are in fact merely spangles of mica. The jade is a species of feldspath combined with a small proportion of talc. Its color varies from a waxen white to a dark green. The Chinese and East Indians make of this stone articles of the most beautiful finish, but very high-priced.

LAPIS-LAZULI.

This mineral is of a fine azure blue color, is usually amorphous, or in rounded masses of a moderate size. It is composed of silica, alumina, soda, and sulphur. It is often marked by yellow spots, or veins of sulphuret of iron, and is much valued for ornamental works.

MALACHITE.

Malachite is an oxide of copper combined with carbonic acid, found in solid masses, of a beautiful velvety green color. It is brought chiefly from Sweden, China, and the Oural mountains. Malachite and grünstein are to be found on all mountains in both hemispheres, but the finest is from the Ukraine.

AMBER.

To account for this singular substance, the ancients invented many charming fables; modern *savants*, though less inclined to poetize what is beyond their comprehension, have been no less puzzled to determine its origin; and though the question is supposed to be solved, it is so merely by conjecture. It is generally supposed to be a resin, derived from an extinct species of pine, and which, long imbedded in the earth, has acquired peculiar qualities. It occurs in round irregular lumps on the margins of the sea, and also in beds of lignite, in clayey schistus, and in calcareous formations. In color, it varies from pale yellow to the rich peculiar hue which bears its name, and to a reddish brown. Though usually diaphanous and even very transparent, it is sometimes almost opaque. Its fracture is perfectly conchoidal, and it is slightly brittle. Its specific gravity is 1·1; a little heavier than water.

It is composed of carbon 79, hydrogen 10·5, and oxygen 10·5. It emits, when burnt, a very fragrant scent.

The quality of negative electricity which amber possesses in a very high degree, had been already remarked in the age of Thales. The phenomenon of a substance to which friction imparted the power of attracting other lighter substances, had so impressed that philosopher that he looked upon it as an animated body. It is from *electron*, the Greek word for this substance, that electricity takes its name.

The Romans set an immense value upon amber. Pliny complains that a higher price was given for exceedingly diminutive human effigies of amber, than for strong and robust living men. It was the fashion for the Roman ladies to carry in the palms of their hands, balls of amber or of rock-crystal. The amber for its delicate perfume, and the crystal for its coolness.

Domitius Nero, among other extravagances, had, in a poem, called the hair of his wife Poppæa, *amber*-colored; this elegant name, given to red hair, brought the color in fashion; and every device was resorted to, even to the wearing of wigs, by the Roman ladies, to give this amber-colored ornament to their heads, in lieu of their own natural and beautiful dark locks.

Shakspeare uses the same epithet as the Roman emperor:—

"Her *amber* hairs for foul have amber quoted."

Very great medicinal properties were formerly attributed to amber, but of this reputation it has been almost entirely despoiled by modern science. It is still worn by children, however, as an antidote to convulsions. An acid called the succinic acid, (from *succinum*, the Latin word for amber,) and an antispasmodic oil, are obtained from it. Amber is also used in the preparation of various perfumes and medical compounds.

It was obtained by the ancients from the coasts of the Baltic Sea, where it is still found, especially between

Königsberg and Memel, in greater abundance than anywhere else in the world. It is there partly cast up by the sea, partly obtained by means of nets, and partly dug out of a bed of bituminous coal. It is obtained in large quantities from the coasts of Sicily and the Adriatic, and is found in different parts of Europe, in Siberia, Greenland, etc. The yellow amber of Dantzic is manufactured to the amount of from 50,000 to 80,000 francs per annum. Transparent yellow amber was formerly in great demand, but, at the present day, the pale tint is preferred. Immense quantities are exported to Turkey, where it is used for pipes, and other ornaments, and consumed in Mahommedan worship at Mecca. Very elegant trinkets, such as necklaces, bracelets, ear-rings, boxes, rosaries, etc., are made of amber. Amber was probably much worn in Shakspeare's day, as he mentions *amber* bracelets, beads, and necklaces.

For personal ornaments, amber had been long out of fashion; but within the last year, it has been worn for demi-toilettes. A necklace, or string of beads, to wind three or four times round the wrist, if the beads are light colored, clear, transparent, and of large size, is now worth in Paris from thirty to sixty francs. Pieces of amber of twelve and thirteen pounds weight have been met with, but such are very rare.

One of the most remarkable circumstances connected with amber is, that it is sometimes found to contain insects of a species no longer extant; leaves, and other vegetable matter, have also been found preserved in amber. The specimens containing these curiosities are highly valued.

"Admire," says Claudien, "the magnificence of the tomb of a vile insect. No sovereign can boast one so splendid—

'Non potuit tumulo nobiliore mori.'"

CHAPTER IV.

PEARLS.

Diversity of Opinions as to the Origin of the Pearl.—Most probable Cause of its Formation.—Oysters inoculated with the Pearl Disease.—Chemical Composition of the Pearl.—Antiquity of the Pearl-Trade.—Ancient Pearl-Fisheries.—Pearl-Fisheries of the Present Day.—Pearl-Divers.

" Rain from the sky
Which turns into pearls as it falls in the sea."

" The liquid drops of tears that you have shed
Shall come again transform'd to orient pearl,
Advantaging their loan with interest
Of ten-times-double gain of happiness."

" Quand une jeune fille agraffe son collier,
Songe-t-elle au plongeur qui pouvait se noyer
En allant cueillir sous le flot qui déferle
Cette fleur de la mer qu'on appelle une perle?"

This singular product of nature puzzled some of the wisest heads of antiquity, and, as usual where a thing is incomprehensible, gave occasion for the wildest conjectures and to the most discordant opinions. Athenæus supposed they were formed in the flesh of the animal, like the hydatides in that of diseased swine; Pliny and Dioscorides, more poetical in their ideas, held them to be a production of the dew; Valentin, that they were the eggs of female oysters; and Samuel Dale, that they were a sort of calculus produced by a hurt received. Admitting that they are the morbid secretions of the animal, there still remain many points unexplained to this day. Redi and M. de Bournon, having opened a number of pearls, invariably found in the interior a foreign body, like a small grain of sand. Supposing that this foreign body introduced in the oyster has caused an irritation, and thence become covered by successive coats of the pearly matter, something like the calculus in the bladder, we can account for the formation of the pearl; though for its shape, its size, and the beauty of its orient, we have no explanation.

According to the most generally-received opinion, pearls are looked upon as a concretion arising from the superabundance of calcareous matter destined to form the shell. In confirmation of this, it has been remarked, that the pearl is always of the same brilliancy, or orient, as the mother-of-pearl that forms the inner coat of the shell. If a pearl is cut, it is found to consist of a succession of laminæ like the onion, and presenting the same appearance in its substance as the shell.

It was the belief that the pearl had its origin in a hurt received by the oyster that suggested to Linnæus the idea of creating the disease in the fresh-water muscle of Sweden, and thus manufacturing pearls at will. The invention appeared of such value to the government, that the great naturalist received a reward of 1,800 dollars—£450, a very large sum in that day and country. Artificial pearl oyster-beds were established in several rivers; but the *modus operandi* by which the pearl was formed, was kept a great secret. It was ascertained, however, that small holes were drilled in the shells of the living oysters, and a foreign substance introduced. In fact, the oyster was inoculated with the pearl disease. Notwithstanding all the trouble taken, the invention disappointed the great expectations founded upon it; the pearls were so small, that the cost was greater than the benefits, and it was finally given up.

The East Indians have a somewhat similar mode of producing artificial pearls, and some of the shells of the large pearl oyster have been found with brass wires inserted throughout their whole length, and the cavities in the interior denoted that pearls had been formed in consequence.

The Chinese throw into a species of shell-fish, (*mytilus cygneus*,) the swan muscle, when it opens, five or six very minute pearl beads strung on a string, and in the course of one year they are found covered with a pearly crust, which perfectly resembles the real pearl.

Several species of bivalved shell-fish produce pearls; but the greater number, the finest, and largest, are produced from the *meleagrina margaritifera lamarck*, a native of the sea and of various coasts. This pearl oyster is in shape not very unlike a common English oyster, but considerably larger, being from eight to ten inches in circumference. The body of the animal is white, fleshy, and glutinous, the inside of the shell (the real mother-of-pearl) is even brighter and more beautiful than the pearl itself; the outside smooth and dark colored.

A considerable number of pearls are also taken from the *unio margaritifera* which inhabits the rivers of Europe; and it is singular, as remarked by Humboldt, that though several species of this genus abound in the rivers of South America, no pearls are ever formed in them.

The pearls are situated either in the body of the oyster, or they lie loose between it and the shell, or lastly, they are fixed to the latter by a kind of neck; and it is said that they do not appear until the animal has reached its fourth year. They have a beautiful lustre, but there is nothing peculiar in their chemical composition, which consists merely of carbonate of lime, and a gelatinous matter. Carthusias asserts that this last substance only enters to the amount of one twenty-fourth in its composition; the other three twenty-fourths are lime and water, of which last there is a large quantity contained in this concretion.

From the composition of the pearl, it is evident it can be easily dissolved in acids; and Cleopatra may well have indulged in the piece of extravagance for which, during so many centuries, she has had the credit, or rather discredit.

The pearl-trade is of the remotest antiquity: We find in history, that from time immemorial, the princes of the East sought this ornament eagerly, and used it on every part of their dress. The Persians, according to Athenæus, sold them by their weight in gold. It is evident that the pearl oyster banks, like those of the common oyster, have not greatly decreased.

Long before the domination of the Persians, the Phœnicians and Babylonians, undeterred by the sandy deserts of Arabia, had found their way to the western coasts, and the islands which favored the trade then carried on with the East Indies, were one cause of this preference. But Ceylon, the ancient Taprobane, and these very coasts, offered also very advantageous products, not the least of which were the pearl-fisheries. According to the author of the *Periplus*, pearls were fished near Manaar, between Ceylon and the main-land. None but condemned criminals were employed as pearl-divers; a practice common to all the nations of the ancient world, who employed them in their mines, in their galleys, in the construction of their public buildings, and execution of all their public works. The pearls were also perforated in the island. This alone proves the fisheries to have been of the remotest antiquity, as the pearl, unless perforated, was then of no use, and this perforation alone, was an operation that required infinite skill.

In the age of Hannon, the above-mentioned author, the chief mart for the pearl trade was the city of Nelkynda, or Neliceram.

Although pearls are ornaments quite as ancient as precious stones, and as indigenous, there is no mention of pearl-fisheries in the Indian authors known to us. Yet the islands and sandbanks between Ceylon and the mainland where these fisheries exist, and where Rama established his famous bridge, during his war with Ravena, are the most celebrated regions in the Indian mythology. It is therefore probable, that our limited knowledge of Hindoo literature is alone in fault, and that so important a fact has not been omitted.

The fable told by Arrien, of Hercules visiting all the Indian Ocean, and finding a pearl with which he adorns his daughter Pandea, is of Indian origin.

As the companions of Alexander mention this fishery, it was evidently of a date anterior to his age.

The pearl-fisheries of the ancients were the above-mentioned, in the Persian Gulf, which, to this day, produce the most beautiful pearls, and in greater abundance than any other place; those of the Indian Ocean, the Red Sea, and the coast of Coromandel. The ancient fisheries in the Red Sea are now either exhausted or neglected; and cities of the greatest celebrity have, in consequence sunk into insignificance or total ruin. Dahalac was the chief port of the pearl-trade on the southern part of the Red Sea, and Suakem on the north; and under the Ptolemies, and even long after, in the time of the caliphs, these were islands whose merchants were princes; but their bustle and glory have long since departed, and they are now thinly inhabited by a race of miserable fishermen.

The principal pearl-fisheries of the present day are those of the coast of Coromandel, on the Sooloo Islands, on the coast of Algiers; those of the banks that extend along the western coast, from the Bahrein Islands, almost to Cape Dsuilfar, and mentioned by Nearchus, are as abundant at the present day as they were in his. In the New World, we have the fisheries of St. Margarita, or Pearl Islands, in the West Indies, and other places on the coast of Columbia.

An inferior description of pearl is found on the Scotch coast.

The oysters fished in the sea near the islands of Karrak and Corgo, contain pearls, said to be of superior shape and orient. They are formed of eight layers or folds, whilst others have only five; but the water is too deep to make the fishing for them either profitable or easy. Besides, the whole profit of the fishery is in the hands of the Sheik of Bushire, who seems to consider the islands as his entire property.

The finest pearls in the world are said to be brought from Bahrein, and constituted at one time the chief trade of the city of Bassora. They are sent to Bassora, where the finest are selected for the European market, and the small ones are sent to China, where they are used as ingredients in medicinal potions, and are sold by weight.

The species of oyster that produces the pearl is found in considerable banks, attached by its byssus to submarine rocks, and always at a considerable depth. One of the largest banks is that which is opposite to Condatchy (Ceylon), some twenty miles long. The reason assigned for these banks remaining as productive up to the present day as they were in the days of the Romans, is the care with which the fishery is conducted. The bank is portioned into seven lots, something as are the coral banks on the coast of Sicily. These lots are worked in succession, one every year. The period of seven years is allowed in order that the animal may attain its full growth, and propagate. Should it be left any longer, however, it is supposed that the pearls would inconvenience the oyster to such a degree, that it would void them. The beds are also properly surveyed, and the state of the oysters ascertained previously to their being let or farmed.

The season for pearl-fishing commences in February, and lasts until about the middle of April. The boats used at the pearl fisheries are from eight to fifteen tons burden, and without decks. The cannon from the neighbouring fort of Arripo gives, at ten o'clock at night, the signal for the departure of the boats assembled in the bay of Condatchy. They leave the shore with a land wind, and proceed to the bank, a distance of from nine to twelve miles. If they reach it before day-light, they anchor close by the government guard-vessel, which is always stationed there, having lights hoisted at night to direct the boats to the banks. The weather is generally calm during the fishing period, the slightest interruption of which is an insurmountable obstacle to the continuance of their pursuit. The men begin to dive as soon as it is light enough to see, and continue their labors till noon, when a gun is fired from the guard-vessel for the diving to cease.

The crew of a boat consists of the tindal, or master, ten divers, and thirteen other men who manage the boat and attend to the divers when fishing. Each boat has five diving-stones, the ten divers relieving each other, so that five are constantly at work during the hours of fishing. The weight of the diving-stones varies from fifteen to twenty-five pounds, according to the size of the diver: some stout men find it necessary to have from four to eight pounds of stone in a waist-belt, to enable them to keep at the bottom, till they have filled their net with oysters. The form of a diving-stone resembles a cone, and is suspended by a double cord. The net is eighteen inches deep, fastened in a hoop eighteen inches in diameter, slung to a single cord.

On preparing to commence fishing, the diver divests himself of all clothing save only a narrow slip of cloth round his loins. After offering up his devotions, he plunges into the water and swims to the diving-stone, which the attendants have hung over the boat's side; he then places the toes of his right foot between the double cord of the diving-stone; the coil of the double rope being passed over a stick, projecting from the side of the boat, he is enabled, by grasping all parts of the rope, to support himself and the stone, and raise or lower the latter for his own convenience, while he remains at the surface; he now puts his left foot on the hoop of the net, and presses it against the diving stone, retaining the cord in his hand, the attendants taking care that the cords are clear for running out of the boat.

As soon as he reaches the bottom, the diver slings his bag round his neck, grasps right and left at all the most promising oysters, and on finding his strength exhausted, makes a signal by pulling the rope in his right hand, withdraws his hold of the rope to which the stone is attached, and which is pulled up after him, and is speedily hauled with his spoils into the boat.

The longer the diver can endure being submerged, the greater of course is his value. Two minutes is considered long to remain under water, though it is said that some men can endure it four, and even five and six minutes. Each diver can repeat the operation as many as forty times in a morning, gathering each time from fifty to four-score shells. They usually stop up their ears with cotton moistened with oil, and also put a few drops in their nostrils and mouth, to prevent the irritation the sea-water would otherwise occasion.

Although severe labor and very exhausting, diving is not considered particularly injurious to the constitution; even old men practise it. The task, under the most favorable circumstances, is a heavy one, but the great and appalling danger attending it is, the risk the divers run of being devoured by sharks. The dread of these sea-monsters would long ago have put an end to the pearl-fishery, but for the superstitious reliance placed by the divers in a set of impostors called shark-conjurors. These charmers pretend to the power of tying the mouth of the shark, or of preventing even his appearance in any particular spot. At sunrise, when the work of the absent divers is just about to commence, the shark-conjurors take their station on the beach muttering their unintelligible charms. This they continue to do until the fleet is in sight on its return, and during all that time they must neither eat nor sleep, their prayers in the case of their so doing being no longer efficacious. To make up for this abstinence they are permitted to drink *ad libitum*; the consequence of which is, that long before the return of the fleet, they are frequently in a state of complete intoxication. Fortunately for the divers, their own wonderful agility in the water, and the vigilance with which their fellows in the boat watch for their slightest signal, are more efficient protection against the attacks of the sea-devils than the muttered gibberish of the conjurors. Busy as the diver is in filling his bag with oysters, he looks sharply out for the shark. If one happens to approach, the diver agitates the bottom so as to render the water sufficiently muddy to obscure the monster's vision, and meanwhile pulls sharply at the rope. At that well-known signal all hands in the boat lend their strength to the rope, and the diver is quickly hauled safely into the boat.

Some of the divers are provided with a charm or amulet, which they wrap up in oil-cloth, perfectly secured from the water, and dive with it on their persons. Others, being Roman Catholics, appear satisfied with an assurance from their clergyman that they have his prayers for their safety; still, strange as it may appear, these Christians have a secret belief in the shark-charmer, and manage to secure his interest. These impostors, who are all Hindoos, are paid by the government, and also receive a perquisite of ten oysters from every boat daily, during the fishery. They have all the resolute audacity of their trade; they maintain their power with the most presumptuous eloquence, they are instantly ready with the most ingenious excuses, and so complete is their ascendancy over the credulity of the divers, that an accident from a shark never awakens the slightest mistrust of the power of these impostors to keep them off

Captain Stewart, in a paper published in the *Transactions of the Royal Asiatic Society*, mentions a circumstance, which strongly characterizes the impudent pretensions of the shark-charmers. He had frequently urged one of them to charm a shark to appear alongside of the vessel in which Captain Stewart then was: but the wily rogue declined doing it, though he positively maintained it to be in his power, on the ground that it would not be right, his business being to keep them away. "During the few days," says Captain Stewart, "that we were employed marking off the ground to be fished, a shark was seen, and reported to me. I instantly sent for the shark-charmer to appear before me, and desired him to account for permitting a shark to appear, at a time when alarm might have a serious influence on the success of the fishery. He replied, that I had frequently requested him to summon a shark to appear, and he had therefore allowed this one the liberty, to please me."

The divers have been bred from infancy to this fatiguing trade. They are mostly Roman Catholics and Hindoos. The most skilful come from Colang, on the coast of Malabar, and from the island of Mandaard. Generally, the owners of the boats, and renters of the oyster-banks, pay the divers certain fixed wages; but sometimes an agreement is made, by which the divers have one-fourth of the produce, and their employers the remaining three-fourths.

The number of oysters procured during the period of the fishery, which is about a month, or six weeks, is prodigious. One boat has been known to bring to land in one day, as many as thirty-three thousand. These are regularly deposited in heaps, as they are brought ashore, until they become putrid, this being necessary in order to remove the pearls easily from the tough matter by which they are surrounded. They are then thrown into large square receptacles enclosed by walls about a foot high, for the better preservation of the pearls. These compartments communicate by four uncovered drains of gradual descent, with a small bath in the centre of the enclosure, so that whatever pearls are swept away by accidental rains, or the washing of the oysters, are carried into this cistern, and none can be lost. Where there are no pavements of the above description, the oysters are heaped on double mats, spread upon the sand within railed enclosures, at the gate of each of which a constant guard is kept for the prevention of thefts. But, notwithstanding all the vigilance that can be used, pilfering prevails through the different scenes of the fishery to a great extent. The divers, the boatmen, the persons employed in washing the oysters and sifting the sand, leave no expedient untried to accomplish frauds. Even the overseers employed to superintend the labors, have been known to attach a viscid substance to the end of their canes, and thus extract from the washing-troughs valuable pearls, with the very instruments used to punish similar delinquents. As the thieves always choose the largest and finest pearls, the loss is heavy. The usual mode of secreting the stolen pearls, is the same as that adopted by the diamond-seekers,—the theft is swallowed; the mode of ascertaining the fact and recovering the property, is also the same as at the diamond-mines,—the suspected delinquent is placed in solitary confinement, and drenched with powerful emetics.

When the oysters are in a state of sufficient decay to be washed, a portion of them is thrown into a canoe, fifteen feet in length, three in breadth, and as many in depth. This canoe is filled with salt water, in which the oysters are allowed to remain for twelve hours, in order that the putrid substance may become perfectly soft, and be thus the more readily disengaged from the maggots, which float upon the surface, and are easily thrown out. From twelve to fifteen naked men are ranged along the sides of the canoe, which is a little elevated at one end, so as to allow the water to run off when it is full. T he oysters are take up one by one, the shells broken from one another, and washed in the water. The stench proceeding from the canoe, during this operation, is the most disgusting that can be imagined. Those employed in it, however, appear wholly unconscious of any disagreeable sensations, and what is stranger still, their health is not at all affected by these dreadful effluvia.

Shells which have pearls adhering to them, are thrown on one side, and afterwards handed to clippers, whose business it is to disengage the pearls from the shells, by means of a forceps and hammer. These pearls, imperfect and deformed as they invariably are, have been generally estimated at forty pagodas per pound weight, and have occasionally reached the price of sixty-four. The part of the pearl that has been adherent must be polished by means of a powder, furnished by the pearls themselves. The roundest and best of them are rendered fit to be strung with other pearls. Many of them are used for setting in brooches and rings. The refuse is mixed with the sand pearl, and sold to make *chunam*, for the palates of certain Chinese epicures, from whom it may be presumed Cleopatra learned the luxury of swallowing pearls.

When all the shells are thrown out of the canoe, the slimy substance of the oysters, turned into mud, remains at the bottom, mixed with sand and small fragments of shells. The dirty water is thrown off in buckets from the lower end of the canoe, and emptied into a sack, hung like a jelly-bag; thus no pearls can escape. Clear water is then poured in at the upper end of the canoe; three or four men stir up the putrescent mass and sand with their hands from the lower end, and prevent the pearls from being washed down. These sink to the bottom, and are kept back by raised pieces of wood, left in hollowing out the canoe. The large pearls are now distinctly seen; the whole of the refuse matter is taken from the canoe, and the bag spread out on a coarse cloth to dry in the sun*.

At certain seasons, the pearl oysters are seen floating

* See Cordiger's *History of Ceylon*.

about on the sea, covering a great extent of surface, and so extremely minute as to appear like the spawn of fish. In this state they are carried by the currents around the coasts of Ceylon, until their increased size causes them to sink, when they form the beds from which so considerable a revenue is derived. The best pearls are generally taken from the most fleshy part of the oyster, near the hinge of the shell; but they are likewise found in all parts of the fish, and adhering to the shell. Some oysters do not contain pearls; others contain one or more. Occasionally an oyster is found containing upwards of a hundred pearls of different sizes, but usually seed pearls.

Though this trade seems at first sight to be one at which prodigious fortunes must be made, it is in truth a very precarious one in which to embark capital. The rent of the bank, and the expenses of fishing it, are enormous and certain, while the produce of a whole boat-load of oysters is frequently insufficient to pay a single diver for his day's work.

During the continuance of the fishing season, there is no spectacle which Ceylon affords more striking to Europeans than the Bay of Condatchy. "This desert and barren spot is at that time," says an eye-witness, "converted into a scene which exceeds in novelty and variety almost anything I have ever witnessed. Several thousands of people, of different colors, countries, castes, and occupations, continually passing and repassing in a busy crowd; the vast numbers of small tents and huts erected on the shore, with the bazaar, or market-place before them; the multitude of boats returning in the afternoon from the pearl-banks, some of them laden with riches; the anxious expecting countenances of the boat-owners while the boats are approaching the shore, and the eagerness and avidity with which they run to them when arrived, in hopes of a rich cargo; the vast numbers of jewellers, merchants, brokers, of all colors and all descriptions, both natives and foreigners, who are occupied in some way or other with the pearls, some separating and sorting them, others weighing and ascertaining their number and value, while others are hawking them about, or drilling and boring them for future use;—all these circumstances tend to impress the mind with the value and importance of that object, which can of itself create this scene*."

The Arabs have a very simple as well as ingenious way of perforating pearls which they intend to string; they take a piece of wood of a porous nature, on the flat surface of which they make small spherical holes with a knife; in these they place the pearls,—which have been counted previous to being confided to them,—so that but a small portion of each goes into the aperture, and then the whole is placed in a little water. The wood swells, and holds the pearl so tightly as to permit of a hole being drilled into the latter by means of a small pointed iron tool, which is turned by means of a wheel. The pearl is loosened, and taken out by letting the wood dry.

The cleaning, rounding, polishing, and sometimes even the perforating and stringing, are performed by negro workmen, who are very skilful in this work.

* Percival.

CHAPTER V.

PEARLS.

Pearl-Fisheries of South America.—Their Value when first established.—Quantity of Pearls brought to Europe.—British Pearl-Fisheries.—Bohemian Pearls.—Large Pearls.—Price of Pearls at the Present Day.—Different Colored Pearls.—Taste of the Orientals.—Goa, the Great Indian Mart for Jewels.—How to preserve Pearls.—Corruptible Nature of the Pearl.—Predilection of the Orientals for Pearls.—Pearls constitute a Portion of Regalia.—Passion of the Romans for Pearls.—When introduced in Rome.—Pearl-Portraits.—Cost of a Pearl Necklace.—Cleopatra outdone.—Cæsar a Connoisseur.—An Enthusiastic Eulogy of the Pearl.—When most worn in France.

After the discovery of America, the traffic in pearls passed, in a great measure, from the East to the shores of the Western World. The first Spaniards who landed in Terra Firma found the savages decked with pearl necklaces and bracelets; and among the civilized inhabitants of Mexico and Peru, they saw pearls of beautiful form as eagerly sought after as in Europe. The new comers lost no time in finding the stations of the oysters, and cities rose into splendour and affluence in their vicinity, all supported by the profits on these sea-born gems. The first city that owed its rise to this cause was New Cadix, in the little island of Cubagna; and the writers of that period discourse eloquently of the first planters, and the luxury they displayed; but now, not a vestige of the city remains, and downs of shifting sand cover the desolate island. The same fate soon overtook the other cities; for, from various causes, and particularly from the indiscriminate destruction of the Meleagrinæ, the banks became exhausted, and towards the end of the sixteenth century, this traffic in pearls had dwindled into insignificance. Of its value when first established, the following extract will afford some notion: "The *quint*, which the king's officers drew from the produce of pearls, amounted to 15,000 ducats; which, according to the value of the metals in those times, and the extensiveness of the contraband trade, might be considered as a very considerable sum. It appears that till 1530, the value of the pearls sent to Europe amounted yearly, on an average, to more than 800,000 piastres. In order to judge of the importance of this branch of commerce to Seville, Toledo, Antwerp, and Genoa, we should recollect that, at the same period, the whole of the mines of America did not furnish two millions of piastres, and that the fleet of Ovando seemed to be of immense wealth, because it contained nearly 2,600 marks of silver. Pearls were so much the more sought after, as the luxury of Asia had been introduced into Europe from two diametrically opposite points; from Constantinople, where the Paleologi wore garments covered with strings of pearls; and from Grenada, the residence of the Moorish kings, who displayed at their court all the luxury of the East. The pearls of the East Indies were preferred to those of the West; but the number of the latter which circulated in the trade, was no less considerable in the times which immediately followed the discovery of America. In Italy, as well as in Spain, the islet of Cubagna became the object of numberless mercantile operations*."

After the conquest of Peru, the number of pearls brought to Europe was, if we believe contemporary writers, immense. Garcilaso de la Viga, in his *Commentaries*, reports that in the year 1564, so fruitful was the fishery, and such the quantity brought to Spain, that they were sold in heaps at public auction in Seville, and that the auctioneer, to get the price raised, offered a bonus of six thousand ducats to the first person who would bid over a certain sum, which

* Humboldt's *Personal Narrative*, vol. ii., pp. 270 and 280.

sum was named. A merchant immediately offered the sum; nor was he over bold in so doing, for he knew, that, however large the amount, such was the quantity and quality of the pearls, even did no one outbid him, he was still a gainer by his bargain. Such was not the case, for another merchant offered more, and the first was fain to content himself with the six thousand ducats he had so easily earned. Nor was the purchaser less pleased, notwithstanding the price he had paid, as the quantity and quality of the pearls ensured him a much larger profit than had obtained the first bidder, who, with a word had won six thousand ducats.

In the year 1587, the reports of the India trade testified to the receipt of eighteen to twenty marcs of pearls of different sizes, but all of transcendental beauty, and of three coffers full of small pearls, sold by the ounce—belonging to the king; for private Portuguese and Spanish merchants there were over thirteen hundred marcs of pearls, besides several bags, which, as belonging to passengers, had not been weighed.

Although Tavernier had not seen the pearl-oyster-fisheries of South America, he gives a very accurate account of them, as they still existed in his day, and mentions five. The first is Cubagna, an island of only three leagues in circumference, and about five distant from the mainland, and one hundred and ten from Santo Domingo. The soil is of the most unproductive nature, and the inhabitants were obliged to bring water from the continent.

This island was renowned throughout Europe for the quantity of pearls it furnished, although the largest did not exceed five carats. We have already spoken of the height of prosperity to which Cubagna rose at one time, and of its present wretched condition.

The next in importance was the fishery of the island of Santa Margarita, or the Isle of Pearls, about a league from Cubagna, which it much exceeds in size. This island, produces every necessary of life, except water, which is brought from the river Cumana, near New Cadix. This fishery was not the most fruitful of the five as to quantity; but the pearls were far superior in size and beauty to those produced by the other deposits. Tavernier says he had in his possession one of these pearls that weighed fifty-five carats, of a beautiful pear shape and fine orient: he sold it to Cha-Est Kan, uncle of the Grand Mogul. One of the finest pearls in the world, the famons Pelegrina, belonging to the crown of Spain, and weighing 250 carats, was fished off Santa Margarita.

The third fishery was at Camogote, near the mainland.

The fourth was in the river La Hacha, along the same coast.

The fifth and last was at Santa Marta, sixty leagues from the river La Hacha. These last three fisheries produced rather large pearls; but generally bad-shaped, and with a leaden hue.

At present, Spanish America furnishes no other pearls for trade than those of the Gulf of Panama, and the mouth of the river La Hacha. The bulk of them, as formerly mentioned, are procured from the Indian Ocean, particularly frem the Bay of Condatchy, in Ceylon.

There are rivers in Britain that produce the pearl oysters, but the produce is so very inferior as to make the fishery of little value. Good pearls have occasionally been found there, and it is said, that one of those in the English crown was taken from the river Conway. It was presented to Catherina, Queen of Charles II., by Her Majesty's chamberlain.

Several pearls of large size have been procured, in the last century, from rivers in the counties of Tyrone and Donegal, in Ireland. One that weighed thirty-six carats was valued at forty pounds; being foul, it lost much of its worth. Other single pearls were sold for four pounds ten shillings, and one even for ten pounds. The last was sold a second time to Lady Glenlealy, who put it into a necklace, and refused eighty pounds for it from the Duchess of Ormond. In his tour in Scotland, in 1769, Mr. Pennant, from whom the above particulars are borrowed, also mentions a considerable pearl fishery in the vicinity of Perth, from which £10,000 worth was sent to London, from 1761 to 1799, but, by the indiscriminate destruction of the mussels, the fishery was soon exhausted.

British, or as they are usually called, Scotch pearls, are seldom large enough or fine enough to be sold singly; sometimes, but very rarely, they attain the size of a pea. They are what is usually denominated brock pearls, and are sold at from five pounds to eight pounds the ounce. They are invariably strung with horse-hair on mother-of-pearl previously formed into the pattern designed, for brooches and ornaments.

The pearls known by the name of Bohemian pearls, are found in the Moldava, from Kruman to below Turenberg.

This river furnishes every year from three hundred to four hundred pearls of the purest water, and very well shaped, besides several hundred imperfect ones. The house of Schwartzenberg owns the greatest part of the banks. The shells that produce the pearls are of a particular species, which it would be advantageous to increase. Besides the Moldava, there is another small river, called the Wattava, which yields a few pearls. They are not fished up, as in the Moldava, from the bed of the river, but taken from the shells thrown upon the banks by the over-flowing of the Wattava. The royal exchequer of Bavaria at one time drew a large portion of its revenues from the pearl fishery of the Iltz, near Passau; but bad management has since completely exhausted it.

European pearls, even when large and well-shaped, differ from the oriental, and even the American, in their lack of brilliancy and orient.

Pearl-dust is used in polishing the pearls and rounding

them as they are in the market; the portion taken from one pearl in the process of cleaning it, serving to round and polish another.

The round pearls are the most admired, but they are generally inferior in size to the oval or pear-shaped ones. One of these mentioned by Tavernier as being the largest, is from the fishery of Catifa. This pearl, reported by Tavernier to be the largest, most perfect, and faultless pearl in the world, had cost the royal purchaser 1,400,000 livres (£50,000 sterling). It measured an inch across, and above an inch and a half in length. Pear-shaped pearls are in great demand for ear-drops.

The price of pearls has much declined in modern times; one cause of this is probably the admirable imitations now made at a very low price in comparison with the real. Jewellers sometimes *make* very large pearls by applying, one on the other, two of the hemispherical rounded tubercles taken from the interior of a pearl-shell.

Some pearls are exceedingly white—such are the most valued; others have a yellowish tint, and others again a leaden hue. The last are South American, and, according to Tavernier, the water in which they are bred, being more full of mud than it is in the East, gives them that color. Tavernier relates that, in the share owned in the Spanish galleons by a famous merchant-jeweller of his acquaintance, there chanced to be six pearls perfectly round, and weighing, one with the other, twelve hundred carats; but all as black as jet. He gave them to the French traveller to be disposed of in the East, but Tavernier could find no purchaser for them.

The yellow tint of some pearls proceeds from their having sojourned too long in the oysters after these have been left to grow putrid. The oysters are allowed to open of themselves, as to force them open with a knife might injure the pearls contained within them. But, when decomposition is allowed to go on too far, the infection changes the color of the pearl. The oysters of the straits of Manaar will open naturally five or six days sooner than those of the Persian Gulf. The heat at Manaar, which is in the tenth degree of latitude north, is far greater than that of the Isle of Bahrein, which is in the twenty-seventh; hence there are few yellow pearls among those that come from Manaar.

The finest South American pearls found their way from the West to the East Indies, the kings and princes of Asia paying higher prices, not only for pearls, but for all precious stones, than could be obtained in Europe.

The Orientals, as Tavernier very naively informs us, share the taste of Europeans with regard to pearls, diamonds, bread,—and women; they give the preference to the whitest.

In all Europe, pearls are sold by the carat-weight like diamonds; but in the East, there are various weights. In Persia pearls are weighed by the *abas;* an abas being one-eighth less than the carat. In the East Indies, and especially in the territories of the Great Mogul and of the King of Golconda and Visapoor, they are weighed by *ratis*, a ratis being also one-eighth less than a carat.

Goa was formerly the greatest mart in all Asia for diamonds, rubies, sapphires, and all kinds of precious stones. Miners and merchants, from all parts of Asia, repaired there to dispose of what they had collected at the mines; as there they were free to sell at their own price or keep their precious wares: whereas, in their own country, the native kings and princes, to whom they exhibited any article of superior beauty, generally put their own estimate upon it, and the owner was compelled to submit. The greatest trade in pearls was also carried on in Goa; not only with regard to those brought from the Persian Gulf, and those fished in the straits of Manaar, on the coast of Ceylon; but it included also those brought from South America.

The great disadvantage attending pearls is that they are liable to alter, especially if they are worn next the skin by certain persons, when they tarnish and lose their brilliancy. One of the methods used to restore their pristine *éclat* was to make pigeons swallow them. But if they remained too long in the crop of the bird, they lost weight. Redi, who tried the experiment, found, after the lapse of twenty hours, they had diminished one-third. The discoloration of these beautiful substances through weather, or other accidental causes, may be prevented by a very simple remedy; pearls kept in dry, common magnesia, in lieu of the cotton wool used in jewel cases, are never known to lose their brilliancy.

The pearl is not like precious stones, incorruptible. Time crumbles them into dust, and annihilates their beauty. On the opening of the tomb in which the daughters of Stilicho had been buried with all their ornaments, after a lapse of eleven hundred and eighteen years, all the riches contained in it were in good condition, with the exception of the pearls, which were so soft that, when pressed between the thumb and finger, they crumbled into fragments.

Pearls have, in all ages, been numbered amongst the most precious productions of the East. The pearl, with its unpretending and *quiet éclat*, its chaste loveliness, the elegant simplicity of its form, that requires but little, if any, aid from art to render it perfect, has been a greater favorite with oriental potentates than even the diamond, with all its eye-dazzling splendour, demanding so imperatively admiration. This taste, which can be traced back in Asia to an epoch far anterior to that of the Persians, has continued undiminished to the present day. A necklace of very large pearls, such as that worn by the Sultan Tippoo, when he died at the gates of his capital, and that which the Shah of Persia wears, is an important part of the royal regalia in the East.

In the West, the taste for pearls did not reach its height until the great Roman empire began to decline; it then became a sort of furor, that led men to make the most costly sacrifices in order to possess these beautiful ornaments.

The victories of Pompey may be said to have first excited a taste for pearls and precious stones in Rome; even as those of L. Scipio and Cn. Manlius had introduced that of chased silver, Attalican stuffs, and table-couches ornamented with bronze; and that of L. Mummius, the fashion of Corinthian bronzes and paintings.

Pliny gives the following account of some of the articles paraded by Pompey before the eyes of the Roman people on the occasion of the third triumph.

"A chess-board, with all its pieces: the board itself made of two precious stones, three feet wide and four feet long (that none may doubt that nature is becoming exhausted, for no precious stones are now to be found of that size.) This chess-board bore a gold moon weighing thirty pounds.

"Three table-couches adorned with pearls.

"A number of gold and silver vases adorned with gems.

"The portrait of Pompey wrought in pearls. Ay, of Pompey! That noble, open brow; that countenance breathing integrity, and inspiring respect to all nations, behold it in pearls! Austerity of customs is conquered, and truly luxury triumphs. Certes, the surname of Great would not long have belonged to Pompey among the men of his time had he triumphed thus on the occasion of his first victory. Thy portrait in pearls, O great Pompey, in that costly superfluity invented for women! In pearls; thou to whom the wearing of them would not have been permitted! Was thy worth thus enhanced! Are not the trophies thou hast erected in the Pyrenees a more truthful image of thee? Truly, this portrait in pearls would be something unworthy and ignominious, if we did not rather see in it a threatening presage of the anger of the gods; and if one did not clearly understand that from that moment this head, loaded with the riches of the East, was shown without the remainder of the body."

It is a pity that, in lieu of wasting his eloquence in denouncing this portrait, Pliny had not entered into particulars, and let future generations know what sort of resemblance pearls could give of a man's features.

Notwithstanding the indignation of Pliny, pearls continued to increase in favor at Rome. Even in his day, he confesses that they were held as the most precious of jewels. He also informs us that the Romans were the only nation that gave the name of *unio* to the pearl; by all others it was called *margarita*. The name of *unio* was given from the difficulty there was in finding two equally white, round, and brilliant.

The women of his day, not content with adorning their sandal-ties with pearls, covered their shoes with them,— "They must even walk on pearls!" exclaims Pliny.

The Emperor Caligula imitated this feminine foppery, and wore pearl-embroidered buskins; and Nero adorned with pearls the sceptres and masks of his players, as well as his couches.

Clodius, to whom his father, the tragic actor, Æsop, had bequeathed an immense fortune, exceeded even the famous prodigality of the Egyptian queen; he not only swallowed, at a banquet, a pearl of immense value, but, declaring that he liked the flavor, gave one to each of his guests to be similarly disposed of. We are not told how many guests there were so favored, or whether, wiser than the rest, one or other did not, by some sleight of hand, reserve the costly dainty for a more appropriate use.

Tertullian tells us a string of pearls was worth one million sesterces (two hundred thousand francs, or eight thousand pounds sterling).

The story of Cleopatra's folly, in dissolving in vinegar a pearl worth £80,729, 3*s.* 4*d.*, has been told for nineteen centuries. The mate of this priceless pearl fell, after the death of the Egyptian queen, into the hands of Agrippa, the favorite of Augustus, who divided it into two equal parts to adorn the ears of the statue of Venus, in the finest temple of Rome, the Pantheon, which he had built. This pearl, even thus divided, was the admiration of Rome.

Modern times record a similar piece of ostentatious folly as having been perpetrated by the princely English merchant, Sir Thomas Gresham. "The Spanish ambassador to the English court having extolled the great riches of the king, his master, and of the grandees of his master, before Queen Elizabeth, Sir Thomas, who was present, told him that the queen had subjects who, at one meal, expended not only as much as the daily revenues of his kingdom, but also of all his grandees; and added, 'this I will prove any day, and lay you a considerable sum on the result.'

"The Spanish ambassador soon afterwards came unawares to the house of Sir Thomas, and dined with him; and finding only an ordinary meal, said 'Well, sir, you have lost your wager.' 'Not at all,' replied Sir Thomas; and this you shall presently see.' He then pulled out a box from his pocket, and taking one of the largest and finest eastern pearls out of it, exhibited it to the ambassador, and then ground it, and drank the powder of it in a glass of wine to the health of his mistress. 'My lord ambassador,' said Sir Thomas, 'you know I have often refused £15,000 for that pearl: have I lost or won?'

'I yield the wager as lost,' said the ambassador, 'and I do not think there are four subjects in the world that would do as much for their sovereign*.'"

* *The History of Banking.* By W. J. Lawson, pp. 24-25.

It is fortunate that subjects do not often show their loyalty, or rather their vanity, by the wanton destruction of these beautiful productions of nature; and notwithstanding his boast, Sir Thomas himself would not have felt inclined to renew the toast in the same costly liquid.

The Romans of both sexes wore pearls in great profusion. In deprecating this effeminacy Pliny becomes eloquent. "What have the waves to do with our garments? That element does not rightly receive us unless we are naked. Grant that there is so great a communion between the sea and the belly, what has the sea to do with the back? It is not enough that our food is procured through perils, if perils are not also encountered for our raiment. Thus, in all that pertains to the body, things acquired at the risk of human life are most pleasing."

Cæsar is said to have undertaken the conquest of Britain from some exaggerated accounts he had heard of the pearls of its coasts, or rather of its rivers. He must have been greatly disappointed if that was his chief inducement, for they were found of a bad color and inferior size. He was himself so great a connoisseur, that he could tell by taking one in his hand what was its exact specific weight and value. The single pearl which he presented to Servilia, the mother of M. Brutus, was worth £48,417 10*s*.

Pierre de Rosnel, in his ***Mercure Indien; ou, le Trésor des Indes***, has the following very enthusiastic, if not very intelligible eulogy of the pearl, proving that in the seventeenth century this ornament was as highly appreciated as it was by the Romans:—"The pearl is a jewel so perfect that its excellent beauty demands the love and esteem of the whole universe. Suidas expresses himself with regard to it thus:—'The possession of the pearl is one of love's greatest delights; the delight of possessing it suffices to feed it (love).' Philostratus, who likewise had the same ideas, in a painting has represented Cupids with garden-shears enriched with pearls; and the ancients were all agreed to dedicate the pearl to Venus. Now to my thinking, the reason for their so doing was, that inasmuch as this goddess of love, the fairest of all divinities, is descended from heaven and is formed of the sea, so in like manner the pearl, the loveliest of all gems. is formed in the sea, and is the offspring of the dew. But he that would be better informed of the excellence and prerogatives of the pearl, let him learn them of the ladies, who will much more relate in its praise than I can write, and will also, doubtless, confess that nothing so well adorns them : the rather as this magnificent jewel has, I know not what fairness the which befits so well the place where they wear it, that it would seem as though nature had therefore intended it."

When philosophers and *savants* indulged in such rhapsodical praise, we can scarcely wonder that the weak-brained majority should have committed extravagant follies to obtain the possession of the world-prized jewels.

In France, pearls have always been in favor, but never, perhaps, so much so as in the reign of Catherine de Medicis and her rival, the celebrated Diane de Poitiers. At the coronation of Marie de Medicis, the dress of that queen and those of the ladies of her court, were covered with pearls. The hair was worn in flowing locks mingled with pearls. Pearls continued to have the preference until the reign of Louis XIV., when diamonds took the lead.

PART FOURTH.

THE QUALITIES, PROPERTIES, AND VIRTUES, OF PRECIOUS STONES.

CHAPTER I.

MARVELLOUS PROPERTIES ATTRIBUTED TO GEMS.

Qualities, Properties, and Virtues, Natural and Supernatural, Physical and Moral attributed in former times to Diamonds and Precious Stones.—Innate Fondness of Man for the Marvellous.—The *Diablerie* of Past Days in Better Taste than that of the Present.—Spirits in Tables, and Angels in Gems.—The Magic of the Diamond.—Opinions of a *Savant* Two Centuries ago.—The Diamond a Peace-maker and a Tale-bearer; an Antidote and a Poison.—The Diamond as an Emblem among the Ancients and the Moderns.—The Gem in the Ephod.—Marvellous Property of the Diamond.

> "Peut-estre que la substance des pierres précieuses, à cause de leur beauté, de leur splendeur et de leur dignité, est propre pour estre le siége et le réceptacle des esprits bons: tout ainsi que le réceptacle des mauvais (selon l'opinion des médicins et théologiens) sont les lieux puants, horribles, affreux, solitaires et les humeurs mélancoliques. . . . . . Comme par ces humeurs les esprits mauvais opérent; pourquoy est-ce que les bons ne pourront pas opérer par les pierres précieuses, et exercer des facultés incroyables, Dieu l'ordonnant et le voulant ainsi?"—Boece de Boot, *Hist. des Pierreries*, liv. ii., chap. iv., p. 158.

> "O! mickle is the powerful grace that lies
> In herbs, plants, stones, and their true qualities."

> "What force the stones, the plants, and metals have to work,
> And divers other things that in the earth do lurk,
> With care I have sought out; with pain I did them prove."

Man has an innate fondness for the marvellous. His imagination leads him to endow whatsoever is rare and precious with supernatural properties, more especially too when the origin of its beauty and rarity is beyond his comprehension. The epoch, which we may term that of the renaissance of the diamond, was also that of the discovery of America and the West Indies. Many of nature's secrets were unveiled. The partial knowledge then gained, far from destroying the illusions of fancy, lent them a fresh incentive, and the wonders disclosed by science were made the basis of new inventions. To real were added imaginary discoveries; hypothesis boldly supplied the links that science yet sought in vain. The treasures with which the Indies enriched Europe, were not looked upon merely as costly and brilliant ornaments; their value was enhanced tenfold by the preternatural qualities ascribed to them.

Nor were these conjectures altogether destitute of foundation. The extraordinary property, invisible to the naked eye, undefinable in its cause but certain in its effects, which had been discovered to reside in the magnet, suggested the belief that other minerals might possess secret and no less wonderful powers. The medical science of that day had borrowed powerful remedies from the mineral kingdom; and preconceived ideas, specious arguments drawn from superficial observation, too implicit a reliance in the poetical fictions by which the ancients explained everything in nature, all combined to endow gems with most wonderful properties, natural and supernatural, physical and metaphysical.

In this age of progress we affect a compassionate interest in the delusions so cherished by our forefathers, and proudly contrast their ignorance with our own superior knowledge. If the truth were told, however, it might be found that we yet cling to what we affect to despise, or at least, look back with regret to the errors we cannot adopt.

Nor is our proud boast of superior enlightenment altogether well-founded. We laugh at the simple credulity of the past; the future will perhaps look back with contempt and derision at our follies. Even while we smile at the poetical fables of the ancients, the sorcerers, magicians, and enchanters of the middle ages, the nuns of London, the Diacre Paris, the counts of St. Germain of the last two centuries, we seek to supply their place with *diablerie* taken from the same source, but infinitely more stupid, certainly less refined, and, in many cases, less innocent. The spirits of the present day are, certes, but mean devils compared with those who called upon our ancestors; and our *mediums* are but wretched dabblers in the cabalistic art of the Agrippas and Great Alberts. The spirits who chose their residence in the most brilliant and beautiful productions of nature, shielding the wearer from pain, shame, sin, and danger, or if their power extended not thus far, warning him at least of the impending evil, these spirits were, it must be confessed, in far better taste than those of the nineteenth century, who take possession of chairs and tables, play idiotic tricks with glass ware and crockery, and require imbecile contrivances to induce them to return answers far more impertinent, mendacious, vague, and unintelligible, than the Delphic oracle's.

Pre-eminent in beauty, brilliancy, purity, strength, the diamond was naturally held to be the most potent in spiritual magic, the most sovereign in its physical effects, of all precious substances.

If, as we suppose, perhaps erroneously, the ancients ignored the art of cutting the diamond, it is not strange that, for personal ornaments, they should have given the preference to pearls and colored gems. But, though they might not value its brilliancy as much as we do, they held it in high esteem as the only unchangeable substance in nature, and one against which iron, and even fire, were powerless.

The perfect gravity with which some learned old authors have discussed the marvellous properties attributed to gems, giving the why and the wherefore,—explaining the seeming phenomena by the beneficent or malignant influence of good or evil spirits; the first *deputed*, the others *permitted*, by God, to work out his own purposes,—rejecting some popular traditions and beliefs, and admitting others no less absurd,—is very amusing. Among those who have philosophized most profoundly on the subject, now and then uttering sound truths, and evidencing the progress made by chemistry and mineralogy in his day, is Anselm Boece de Boot, physician to the emperors Rodolphe II. and Maximilian II., and who wrote in 1664. Unfortunately for the generality of curious readers, the learned man's statements and commentaries are enveloped in such a farrago of words, that it is a labor of patience to wade through them. We will take the liberty to disencumber some of his gems, for the benefit of those who do not care to go to the trouble of extracting them from his treatise.

The savant does not deny the supernatural effects attributed to gems, but explains them as not in the nature of the stone itself, but as being imparted to it by supernatural influences:—

"The supernatural and acting cause is God, the Good Angel and the Evil One; the Good, by the will of God, the Evil, by the permission of God." And elsewhere the Doctor adds:—"But that which God can of Himself perform, that may He likewise do through his ministers and good angels, the which, by a special grace of God, and to secure the preservation of things, are enabled to enter precious stones. And even in this fashion is it given unto them to guard men from peril, and do them some singular favor. But, even as we can nothing assert with certainty touching the presence of angels in precious stones, so neither should we believe too implicitly in precious stones, or ascribe too much unto them. For my own part, I do rather incline to the opinion that the Evil Spirit, taking the semblance of an Angel of Light, taketh up its abode in precious stones and enacts by them prodigies, in order that, instead of having recourse to God, we may rest our faith on the said stones, and consult them rather than God, when we would compass some object. Thus perchance are we deceived in the turkois by the Spirit of Evil."—B. i., c. xxv., p. 3.

In the foregoing extract, we see that the learned man was not at all in doubt as to the existence of spirits in gems, but only as to whether such spirits were angels or demons.

Farther on, he tells us, that, "from their purity, beauty, and brilliancy, it is most probable gems were selected as receptacles for good spirits, even as filthy, stinking, and frightful places are usually the abodes chosen for evil and unclean spirits."—B. i., c. xxv., p. 158.

From the extreme brilliancy of the diamond, and its purity, it was consecrated to all that was celestial, and was accordingly supposed to triumph over all means employed to subdue it, the solar ray excepted. It was reputed a powerful talisman, and when under the planet Mars, productive of great success.

For many ages, it was held to be powerful against poisons, plagues, sorcery, enchantments, insanity, vain terrors, evil-spirits, and the nightmare. It would sweat in the presence of poisons. It was esteemed a safeguard to virtue.

The diamond has been despoiled of all these properties by the moderns; as to the last, so completely has the gem changed its nature, that it has now the effect of corrupting rather than of preserving virtue.

Yet, however beneficial the effect, as a preventive when worn, it had quite a contrary one when taken internally. Diamond powder, swallowed, had a property so venomous, that nothing could save the victim. The disciples of Theophrastus Paracelsus, the chemist, asserted that such had been the cause of his death; this was probably said to conceal the error into which their master had fallen; who, having predicted that his life would be prolonged far beyond the usual term, by means of his elixir, yet died in the flower of his age.

The diamond bestowed victory, fortitude, and strength of mind. It calmed anger, and strengthened wedded love; hence, it was called the stone of reconciliation. That the diamond still deserves this name, and is often made a peace-offering to heal imaginary or real heart-wounds inflicted on the vanity or the self-love of some afflicted fair, there is no doubt whatever.

But if the beautiful gem was, and yet is, frequently a peace-maker, we have right reverend authority that it sometimes caused sad mischief, by its indiscreet revelations. The sure test of conjugal truth, was to place without her knowledge, a diamond beneath the married dame's pillow. The lady's dreams proclaimed aloud the state of her heart,—

> "And mutters she in her unrest
> A name she dare not breathe by day."

The poet neglected to ascertain if Parisina's jealous lord thus tried the property of the gem.

The learned De Boot, in allusion to this virtue of the diamond, takes occasion to discourse at some length, as to whether the power of discrimination between right and wrong, legal and illegal affection, be a natural quality of the stone, or belongs to a spirit residing in it. The very extraordinary arguments he uses, being couched in words too crude for modern taste, we refer the reader to his treatise. B. ii., c. iv., p. 154.

Though the diamond was not believed to be fusible by fire, the splendour of its lustre and its properties were supposed to be affected by heat; therefore Wolfgangus Gabelschoverus advises those who wear these gems, to take their ring from their finger at night, and put it in a cup of cold water, or on a marble, or other cool surface.

Some very marvellous statements are made by De Boot, on the authority of one of his friends, a physician also, who asserted that he possessed a secret by means of which he could, with the greatest ease, place a diamond on the point of a needle; and also, by the help of his nails only, and without making use of any instrument, divide a diamond into scales like those of the specular stone.

Among the ancients, the diamond was a symbol of severe and inexorable justice, and of the impassibility of fate. Hence, the judges of Hades were described as having hearts and bosoms of adamant. Of adamant also were the clouds in which destiny was shrouded, to show its irrevocable, immutable nature.

The diamond has been used as a symbol, in more modern times, to signify innocence, constancy, faith, strength, etc., etc. Many princes, and among them Cosmo de Medicis, made use of it thus.

A Jewish legend relates of the gem,—supposed to have been a diamond—worn on the ephod by Aaron, the high priest, that when a man was charged with a crime, if he was guilty, it became dark and dim: if the accused was innocent, it sparkled with increased lustre. If the sins of the Hebrews were to be punished with death, the gem assumed a bloody hue.

But all these wonderful virtues of the diamond are eclipsed and thrown into the shade, by one most marvellous and unique, for it was attributed to no other gem—the faculty of multiplying its species!

Boetius de Boot, quoting from another learned man,

relates that a lady of good family had two hereditary diamonds, which produced several others, and thus left a posterity. The comments of the narrator, on this generative faculty, are no less curious than the statement itself. He does not inform us whether these descendants were born small, and grew in size from infancy to maturity.

According to Pliny, there existed between the diamond and the magnet a natural antipathy. "There is," says he, "such a disagreement between a diamond and a loadstone, that it will not suffer the iron to be attracted; or if the loadstone be put to it, and take hold of it, it will pull it away." It is needless to observe that no such antipathy exists.

CHAPTER II.

OF THE VIRTUES AND PROPERTIES, SPIRITUAL AND PHYSICAL, OF PRECIOUS STONES.

Precious Stones used medicinally.—The Five Precious Fragments.—All Gems averse to Poisons.—Gemmed Cup given by Louis XI. to his Brother.—Talisman of the Count of St. Pol.—Ring sent to Queen Elizabeth.—Innate Properties of Gems strengthened by Magic.—Ring of Louis, Duke of Orleans.—Properties of the Ruby.—Death-presaging Ring of a German Philosopher.—Male and Female Carbuncles.—Properties of the Sapphire.—Consecrated to Phœbus.—Tables of the Law.—Sapphires.—Male and Female Sapphires.—Properties of the Emerald.—The Eye-glass of a Roman Dandy.—Properties of the Topaz—of the Amethyst—of the opal—of the Turkois—of the Beryl—of the Agate—of the Jasper—of the Heliotrope.—Great Reputation of Coral.—Consecrated to Jupiter and Apollo.—Amber.—Pearls.

"For it is said, and hath been said full yore,
The emeraud greene, of parfite chastitie,
Stole ones away may not recovered be."
CHAUCER.

"The verdant, gay green smaragdus,
Most sovereign over passion."
DRAYTON.

EVERY precious stone has its special virtue, though some properties might be said to belong to all.

The more precious the stone, the more powerful were its virtues. The ancient pharmacopœia counted among its most sovereign remedies, a very costly compound called the "Five Precious Fragments," consisting of powdered rubies, topazes, emeralds, sapphires, and hyacinths. As the patients were not likely to analyse the beverage, it is probable, that in the generality of cases, they swallowed false gems, while the physician or chemist wisely pocketed the real, which thus escaped a barbarous and useless destruction. Antiquity and the middle-ages believed as implicitly in the influence of stones and plants, as in that of the heavenly bodies, and it will be seen, that they associated these influences, to give them greater power. Doubtless, the belief in the efficacy of gems was frequently corroborated by the results attending the use made of them; but, in such cases imagination had a large share. The entire faith, the hope, the certainty entertained of a cure, especially when the patient was of a nervous, impressionable temperament, or the disease was one of mind, constituted a powerful aid to medical science. Hope, in itself, is nature's great physician, and imagination works out realities.

There is an oriental tradition, that Abraham wore a precious stone round his neck, which preserved him from disease, and which cured sickness when looked upon. When the patriarch died, God placed this stone in the sun: hence the Hebrew proverb: "When the sun rises the disease will abate."

All gems were supposed to express an antipathy to poisons. Holinshed tells us, that King John, observing a moisture on some precious stones which he wore, thought it an indication that some pears he was about to eat, contained poison.

The day after the interview of reconciliation which took place between Louis XI. and his brother the Duke of Guienne, the king sent to his brother, bidding him accept as a token of fast friendship, a beautiful golden cup, enriched with precious stones which were gifted with the power of preserving from poison whosoever used it.

When the Constable, Count de St. Pol, was led to the scaffold, he drew from his finger a gold ring, enriched with diamonds, and begging the chaplain to place it on the finger of the statue of Our Lady, "Father," added he, "here is a gem I have always worn about my neck, and which I have highly valued, for it has a marvellous virtue:

it preserves the wearer from the plague and every other infection. Bear it, I pray you, to my grandson, Louis, and tell him I desire he will always keep it for my sake." Louis XI. kept the stone himself.

Fosbroke says, that Lord Chancellor Hatton sent to Queen Elizabeth a ring against infections, to be worn in the bosom.

Sometimes the virtue of the stone was enhanced by sorcery.

When John the Fearless, Duke of Burgundy, caused a justification of his base murder of his cousin, the Duke of Orleans, to be read and published, among the accusations he brought against the murdered prince, was that of having had dealings with the devil, to destroy the life of the king. The whole of the magical process by which a sword, dagger, and ring had been charmed for the use of the Duke of Orleans, was described at length. By the virtue of the ring, which had been charmed in the name of the false goddess Venus, the Duke could fascinate and bend to his will any woman. The young and gallant Louis of Orleans needed no such aid from Satan. Nature had given him the all powerful magic of a handsome person and winning tongue.

The oriental ruby, or carbuncle of the ancients, either worn as an ornament, or reduced to powder and taken internally, was an antidote to poison, and preserved from the plague; it banished sadness, repressed sensuality, put evil thoughts to flight, dispelled fearful dreams, diverted the mind, and guarded against all illness. If misfortune threatened the wearer, it gave warning by a change in its color, which darkened greatly; but when the evil or peril was no longer to be feared, it resumed its primitive bright hue.

In confirmation of the above, we have the following occurrence related by Wolfgangus Gabelschoverus: "This have I often heard from celebrated men of high estate, and also know I it, woe is me! from my own experience. For, on the 5th day of December, 1600, after the birth of Christ Jesus, as I was going with my beloved wife, Catherina Adelmannie (of pious memory) from Stutgard to Caluna, I observed by the way that a very fine ruby which I wore mounted in a gold ring, (the which she had given to me) lost repeatedly, and each time almost completely, its splendid color, and that it assumed a sombre blackish hue, which blackness lasted not one day but several; so much so that being greatly astonished, I drew the ring from my finger and put it into a casket. I also warned my wife that some evil followed her or me, the which I augured from the change in the ruby. And truly I was not deceived, for within a few days she was taken mortally sick. After her death the ruby resumed its pristine color and brilliancy."

Though the ruby did such good service to man, it had its evil effects also, inasmuch as it shortened the sleep of the wearer, agitated and disturbed the circulation of the blood, inclining him to anger.

The ruby, formerly called the carbuncle, was held in great repute as being supposed to give light in the dark. Hence the names of *anthrax*, *pyrope*, *coal*, given to it by the ancients. Garcias ab Horto, physician to the Viceroy of the Indies, relates that he had heard persons assert, they had witnessed this property of the ruby, but adds, *that he himself believed it not.*

Louis Verolam relates that the King of Pegu wears carbuncles of such a size and lustre, that whosoever looks at the king in the dark, sees him as resplendent as though he were illumined by the sun; but neither had he seen this wonder.

This splendour of the ruby in the absence of light is, up to a certain point, confirmed by modern writers. A recent traveller tells us, that at a Siamese court, where only a subdued light is admitted, the diamonds and carbuncles on the king's person glittered and flashed like miniature lightning.

But the ancients were not content with the glorious beauty of nature, and invariably added marvels of their own coining. Ælian, in his eighth *Book of Animals*, narrates that a woman of the name of Heraclea, having cured a stork of a broken leg, the grateful bird brought, and dropped into her bosom, a carbuncle, the true amethyst of the Ethiopians, which shone in the darkness of night like a lighted lamp.

Pliny tells us that there are male and female carbuncles —the males being more acrid and vigorous, the females more languishing.

The sapphire, when worn by an impure, intemperate, and debauched person, lost its lustre and beauty. It had this property, in common with the majority of precious stones, which thus manifested their abhorrence of vice and impurity. It inspired continence, repressed loose thoughts, cured diseases of the skin, etc., etc.

Placed on the brow it stopped hemorrhage. To look often at a sapphire was excellent to preserve the eye-sight. Powder of sapphires made into a pill and placed on the eyes, drew out any dust, insect, or other foreign substance that might have fallen into them, and cured them if inflamed or irritated by small-pox, or other diseases. But the application was to be daily renewed for some time. The powder was also taken in milk for internal diseases, and was then esteemed sovereign against plagues, fevers, poisons, hysteria, etc.

So great was the power of the sapphire on *venomous* creatures, that if it was placed over the mouth of a phial containing a spider, the insect died on the instant.

Placed on the heart, it cured fever, and bestowed strength and energy. Its power to inspire pure and chaste thoughts caused it to be recommended to be worn by ecclesiastics. St. Jerome, in his explanation of the 19th chap. of the prophecy of Isaiah, asserts that the sapphire conciliates to the wearer the favor of princes, calms the fury of his enemies, dispels enchantments, delivers from prison, and softens the ire of God.

The ancients held the sapphire in the highest honor: in the sacrifices of Phœbus, his worshippers, to propitiate the god, offered him a sapphire.

Epiphanes states, that the vision which appeared to Moses on the Mount, was in a sapphire, and that the first tables of the law, given by God to Moses, were of sapphire.

Lapidaries designated the deep blue sapphire as the male, and the pale blue as the female.

The emerald worn as an amulet about the neck, or set in a ring, put evil spirits to flight, and was a preserver of chastity. *Per contra*, it betrayed inconstancy by crumbling into fragments, when it could not prevent the evil. It testified in the same way its sense of incapacity to subdue illness. It taught the knowledge of secrets and of future events; it bestowed eloquence, and increased wealth. This superstition suggested to one of the most charming English poets, the following beautiful lines:—

"It is a gem which hath the power to show,
If plighted lovers keep their faith or no:
If faithful, it is like the leaves of spring;
If faithless, like those leaves when withering.

"Take back again your emerald gem,
There is no colour in the stone;
It might have graced a diadem,
But now its hue and light are gone.

"Take back your gift and give me mine—
The kiss that sealed our last love vow;
Ah! other lips have been on thine—
My kiss is lost and sullied now!

"The gem is pale, the kiss forgot,
And more than either you are changed;
But my true love has altered not,
My heart is broken, not estranged*!"

* Miss Landon.

"He who dreams of green gems, (*prasini coloris,*) will become renowned, and meet with truth and fidelity*."

The falling of an emerald from its setting has been held as an ill omen to the wearer, even in modern times. When George III. was crowned, a large emerald fell from his diadem. America was lost during his reign. It is probable, however, that the fallen stone was picked up and refitted in the crown,—an event not likely to happen in the case of the other gem.

Pliny asserts that the emerald gives forth an exceedingly brilliant light; and, in corroboration, tells a story about the tomb of a petty king called Hermias, in the island of Cyprus, near the fisheries. On this tomb was a marble lion with emerald eyes. Such was the extraordinary brilliancy of the emeralds, and so far out at sea did it extend, that the frightened fish fled to a great distance. The fishermen having ascertained the cause of the scarcity of their prey, removed the emeralds, and substituted less costly eyes. The gems did not frighten *them*, knowing fellows! though they did alarm the silly fish.

Objects were supposed to appear in a more favorable light, when seen through emeralds. Nero used an emerald to look at the combats of the gladiators. But Nero was something of an exquisite, and may be suspected of having used the gem in that way as a bit of foppery, rather than for any real advantage he derived from it. Even now, many *lionnes* would think objects far more interesting viewed through such a medium.

Besides its metaphysical properties, the emerald had very powerful medicinal virtues, being, like all precious stones, "of a cold, dry nature." Taken internally as a powder, it was considered a cure for venomous bites, pestilential fevers, and many other diseases. When hung round a child's neck, it guarded from epilepsy. If powerless to

* *Achmet Seirim*, c. 247.

prevent or cure any evil, it shivers into atoms, "being, as it were, bound to expel the evil, or confess itself vanquished in the combat it sustains*." Applied to the lips, it stopped hemorrhage; worn round the neck, it dispelled vain terrors, put to flight evil spirits, and cured fevers. It was a restorer of sight and memory. Our oft-quoted authority, De Boot, gives the method of extracting from emeralds the *coloring matter*, which, taken internally, was considered so efficacious.

To the topaz, the chrysolite of the ancients, was attributed the same quality ascribed to the ruby, of giving light in the dark. A topaz was said to have been given to Monsieur Adelbert, tutelary president of the Egmondeuses, by the noble lady Hildegarde, wife of Theoderic, Count of Holland, the which gem threw out at night so resplendent a light, that in the chapel where it was kept prayers were read without need of a lamp.

The topaz lost its color in the presence of poisons. It dispelled enchantments, if bound on the left arm, or hung round the neck, set in gold. Worn on the left hand, it preserved from sensuality. It calmed anger and frenzy. It dispelled night-terrors, banished melancholy, strengthened the intellect, cured of cowardice, and brightened the wit.

Its medicinal virtues were great. Powdered and taken in wine, it cured asthmas, want of sleep, and divers other maladies.

The amethyst diverted evil thoughts, made the wearer diligent, and procured him the favor of princes. Its chief virtue was that it preserved its wearer from intemperance, and acted as a cure for inebriety.

The chief virtue of the hyacinth was, when worn as an armlet, to preserve its wearer from the plague. It also

* "Elle doit ou lever le mal ou céder comme s'avouant vaincue par le plus fort dans le combat qu'elle rend." B. de Boot: *Traité des Pierreries.* L. ii., ch. liii., p. 253.

inclined to sleep, increased riches, honors, prudence, wisdom, and was a safeguard against lightning.

The opal was called by the ancients "pederos," from "puer" a child, inasmuch as, like a fair and innocent child, it was worthy all love.

Even as the opal united in itself the colors of every other gem, so also was it supposed to possess all their qualities both moral and healing. It was especially good for the eye-sight, keeping it clear and strong.

The turkois protects, by drawing upon itself the evil that threatens its wearer; but this property belongs only to the turkois that has been given, not to a purchased one. The following extract from Boetius de Boot shows the faith once had in this gem :—

"It" (a turkois) "had been thirty years in the possession of a Spaniard, who resided within a short distance of my father's house. After his death, his furniture and effects were—as is the custom with us—exposed for sale. Among other articles was this turkois; but, although many persons, admirers of its extraordinary beauty during its late master's life, were now come to buy it, no one would offer for it, so entirely had it lost its color. In fact, it was more like a malachite than a turkois. My father and brother, who had also gone with the intention of purchasing it, being well acquainted with its perfections, were amazed with the change. My father bought it notwithstanding, being thereunto induced by the low price put upon it. On his return home, however, ashamed to wear so mean-looking a gem, he gave it to me, saying: 'Son, as the virtues of the turkois are said to exist only when the stone has been given, I will try its efficacy by bestowing it upon thee.' Little appreciating the gift, I had my arms engraved upon it as though it had been an agate, or other less precious stone, such as are used for seals and not ornaments. I had not worn it a month before it resumed its pristine beauty, and daily seemed to increase in splendour."

The doctor goes on to state the various accidents from which his gem guarded him at the cost of its beauty and even value, for it broke at last. But he had the fragments mounted, and they severally possessed the virtues of the entire stone. This sympathetic property of the turkois, manifested by a change of color, is alluded to by several English poets. Thus, the flatterers of Sejanus, in Ben Jonson's play of that name,—

> " Observe him as his watch observes his clock,
> And true as turkois in the dear lord's ring
> Look well or ill with him."

And again Donne :—

> " As a compassionate turkois that doth tell,
> By looking pale, the wearer is not well."

Two or three centuries ago, no gentleman thought his hand adorned, unless he wore a fine turkois ; but the gem was less patronized by ladies. It was esteemed the noblest and most valuable of opaque stones.

The turkois relieved or prevented head-aches, reconciled lovers, and appeased hatred.

The beryl was a protection against the snares of enemies. It was efficacious in liver complaints, hysteria, jaundice, convulsions, in diseases of the mouth, throat, and face. When powdered, it cured eyes that were sore, wounded, or contused. The magi held it to be a sovereign remedy against idleness, a sharpener of the intellect, and a reconciler of married people. The beryl, or aigue-marine, rendered the wearer successful in navigation, and preserved him from all danger, however rough and tedious the voyage.

Of the onyx, we have not quite so good a character to give as we have been able to furnish other gems with. If worn on the neck, it excited spleen, melancholy, vain terrors, and other mental perturbations; all of which were, however, counteracted or cured by the presence of the sardonyx, or cornelian. The latter gem was also efficacious against infections, pestilential humors, hemorrhage—for the cure of these maladies it was necessary to take the stone in powder. The cornelian was also excellent to clean and whiten the teeth. Cardan asserts that it won law-suits and enriched the wearer.

The agate, whether a cornelian, a sardonyx, an onyx, or any other stone coming under the general denomination of agate, has the property of preserving from the bite of venomous animals, and particularly that of the scorpion. It appeases pain, and if hung round the neck, so that it rests on the neck, subdues amorous thoughts. The Persians believed that its *scent* turned away tempests and arrested the impetuosity of torrents. If the stone be of a single color, it renders the wearer invincible. This last virtue is especially attributed to the chalcedonyx; and it was said that Milo, of Crotona, was indebted to one for the increase of his wonderful strength, always wearing it when about to undertake one of his feats. It is probable he forgot his talisman when he tried his last exploit.

The blood-red jasper was effective to stop hemorrhage, if worn or applied to the wound.

The heliotrope had the property, according to the Great Albert, of rendering its wearer invisible.

Coral was formerly in great repute; we have many grave authorities in favor of its various virtues. That its fame was somewhat on the wane in Pliny's time, we have from that author's remark, that " formerly it was deemed excellent as an antidote to poison." In the middle-ages, however, it had regained a high place in the pharmacopœia, and also as an amulet. Brand quotes from an old work the following testimonial in its praise: " Witches tell that this stone withstondeth lyghtenyng, and putteth it as well as whirlewindes, tempestes, and stormes, from shippes and houses that it is in." Pierre de Rosnel tells us, that coral worn by a healthy man will be of a handsomer, more lively red, than if worn by a woman. It becomes pale and livid, if worn by a person that is ill and in danger of death. Thus, in the *Three Ladies of London* (1584)—" You may say jet will take up a straw, amber will make one fat, coral will look pale when you be sick, and crystal will staunch blood."

Coral was a talisman against enchantments, witchcraft, venom, epilepsy, assaults of Satan, thunder, marine tempests, and other perils. On account of these beneficial properties, it was consecrated to Jupiter and to Phœbus. Hung round the neck, it stopped hemorrhage. Taken internally, it was esteemed an excellent remedy. Bocce de Boot says, he was himself cured of a dangerous pestilential fever by taking six drops of tincture of coral. Armand de Villeneuve asserts, that ten grains of coral given to an infant in its mother's milk, provided it be a first child, and this be its first food, will effectually preserve it from epileptic, or any other fits, throughout life.

In modern nurseries, the faith in the coral and bells suspended round the necks of infants still exists, for the red coral repels witchcraft, and bells were originally used to scare away evil spirits.

> "So children cutting teeth receive a coral."
>
> BYRON.

Amber, like coral, has preserved among the moderns, to a certain extent, its reputation for talismanic powers. It is still worn by children as a preservative against many diseases.

The pearl had great medicinal virtue when taken inwardly ; but no influence on passions or events when worn. The oneirocritics, or interpreters of dreams, drew interpretations from pearls.

Among the dregs administered to restore to reason the unfortunate Charles VI., a decoction of pearls and distilled water was given to him. Large quantities of seed pearls are used, even at the present day, throughout Asia, in the composition of majoons, or electuaries, to form which all kinds of precious stones are occasionally mixed, after being pounded. Diamonds are never used as internal medicine, being considered, on account of their hardness, indigestible. The majoon, in which there is a large quantity of pearls, is much sought for and valued for its supposed stimulating and restorative qualities.

Margaritæ significant lachrymarum flumen. Pearls signify a torrent of tears*.

We have spoken only of the merits and properties of stones used as ornaments. To speak of the virtues of the innumerable stones employed in the ancient pharmacopœia, would be foreign to our subject. Even in mentioning but those of the gems worn, we might fear to have been prolix, but that we have furnished our fair readers with excellent apologies for forming collections. The more jewels, the more guardian spirits; and surely, very safe may be deemed the fair one whose form is encompassed by angels.

* Astrampsychins.

CHAPTER III.

GLYPTICS.

Intaglios.—Cameos.—Antiquity of the Glyptic Art.—Graven Gems in the Ephod and Pectoral.—Graven Talismans.—Advice of a *Savant*.—Stones preferred by Ancient Artists.—Celebrated Antiques.—Shell-Cameos.

The art of engraving on stones was practised in Egypt forty centuries ago. It is probable that the graven gems that adorned the ephod and pectoral of the high priest of the Hebrews were the work of Egyptian artists, who, escaping from the tyranny of Pharaoh, had followed Moses into the desert. Though a sojourn of four centuries in Egypt must have made the Hebrews acquainted with many of the handicrafts of the country,t hey had been employed in ruder labor too long to be very skilful in the arts that supplied the luxuries required by their elegant and refined masters.

So great was the reliance formerly placed in precious stones, both as talismans and for their reputed healing properties, that immense sums were expended to procure these expensive remedies. Nor has that reliance long ceased to exist. Not half a century has elapsed since it was yet customary to apply to the wealthy for the loan of rings in which precious stones were set, in order to place them on the part of the body that was affected by the disease the gem was supposed to cure. When the jewel was to be held in the mouth, it was secured by a string to prevent its slipping into the throat.

In order to increase the power attributed to gems, certain figures or characters were graven upon them at the precise moment when the conjunction of the heavenly bodies was favorable to the object in view. Hence, if the gem was intended to render its wearer victorious, the effigy of Mars, or that of Hercules subduing the Hydra, was graven upon it at the precise hour when the aspect of the heavens was indicative of victory. Here again the wise De Boot has reflections of such weight, we cannot refrain from quoting them:—

"In good sooth, I am fain to confess that supernatural effects are after this fashion produced, God having permitted it should be so. But, as I have already said, this is done by means of evil spirits, who take up their abode in the substance of the precious stones, constrained thereunto by the vain credulousness of man, and by a pagan impiousness; taking undue advantage of the stone, to the end that they may conceal or annihilate its natural faculties, rendering them unrecognizable, substituting in their place false ones; and by these means leading man to vanities and superstitions, making him forsake the true worship of God, subjecting him to their will, and losing his soul to all eternity. Those, therefore, who would attract good spirits to inhabit their gems and benefit by their presence in them, let them have the martyrdom of Our Saviour, the actions of his life which teach virtue by example, graven upon their jewels; and let them oft contemplate them piously; without doubt, with the grace of God and the assistance of good spirits, they will find that not in the stone only, or the graven image, but from God are its admirable properties."

The above arguments prove the learned doctor was no less pious than credulous.

Ancient engravers frequently chose the stones most in accordance with the subject they intended to represent: thus they engraved the figure of Proserpine on a black stone, Neptune and the Tritons on the beryl, Bacchus on the amethyst, the story of Marsyas on red jasper. Some cameos were cut on stones of a single color; fine ones of this kind are to be seen in the Bibliothèque Impériale of Paris, among others the head of Maximilian III. on an agate, and Ulysses on a cornelian. But the majority of cameos were cut on sardonyxes or agate-onyxes, that is, stones exhibiting two, three, and even four layers or bands of distinct colors. In the latter case, it was necessary that the artist should be not only a skilful designer and modeller, and be well acquainted with the mechanism of the glyptic art, but it was requisite that he should also possess a special genius, a peculiar tact, to take advantage of the different colors of the stones, use them in their proper places, and adapt them to the divers objects he intended to represent.

Where stones of various colors are employed, the figures are usually cut on the white portion, while the brown or black is the ground. When there are three or four layers of different colors, the engraver is enabled to vary those of the hair, beard, and draperies, so as to produce the most pleasing effect. Beautiful specimens of this description are in the Bibliothèque Impériale. One of the finest is the apotheosis of Augustus, a tricolored sardonyx representing twenty-two figures. This superb antique is sometimes called the Agate of the Sainte Chapelle, and was brought from the East in the reign of Saint Louis. Given by Charles V. to the Holy Chapel of his palace, it was there supposed to represent the triumph of Joseph over Pharaoh. Another fine cameo represents the apotheosis of Germanicus, a three-colored sardonyx. This fine antique was brought from Constantinople, by Cardinal Humbert, under the pontificate of Leo IX., and then given to the Benedictine monks of Toul. The eagle which bore the young prince led to the belief, that this cameo represented the apostle John. Criticism of a higher order having dispelled this error, the monks presented the antique to the King, in 1684. Ceres conducting Triptolemus in her chariot drawn by two dragons, is another splendid cameo. The contest between Neptune and Minerva, during which the divinities present the city of Athens with the horse and olive tree, is engraved upon a much smaller sardonyx, but the workmanship is beautiful. Another fine cameo represents Silenus teaching a group of cupids. A Jupiter grasping in one hand his thunderbolts, and in the other his sceptre, and having his eagle at his feet, is a superb tricolored sardonyx. In the imperial cabinet of Vienna, at La Haye, and other places, there are also magnificent cameos.

Cameos were used chiefly for dress ornaments, clasps, etc., while intaglios, being intended for seals, were set in rings. The latter were much more common than the former. One of the finest known is the aigue-marine that bears the name of the engraver Evodos, and represents the head of Julia, daughter of Titus: it is now in the Bibliothèque Impériale, and was formerly thought to be a head of the Virgin.

The ignorance and barbarity of the dark ages, succeeding to the refinement and the learning of the art-worshipping Greeks and voluptuous Romans, the relics of the fine arts that survived the universal destruction were preserved in cloisters and churches; changing, not their nature, but their destination: from having been emblems of the poetical myths of an extinct faith, they became the ornaments of the insignia of a new creed.

In accordance with the taste of their day, the crusaders, on their return from the Holy Land, sanctified the fruits of their rapine and pillage by dedicating them to God; the ancient cameos, the intaglios, and the other precious gems of which the emperors of the East had spoiled Rome, returned to the West, and were deposited in the chapels and basilicas of Europe. Such was the origin of the famous Venetian treasure of St. Mark. Numbers of these precious mementos were more indebted for their preservation to the ignorance of their new possessors than to their beauty and real value, being looked upon as subjects taken from Holy Writ; and this ignorance gave rise, in some cases, to the most ludicrous misapplication and misinterpretation.

With the renaissance of other arts returned the taste for cameos and intaglios, and the study of the glyptic was resumed. The Medici family contributed powerfully to its development; and to such perfection did it arrive through their encouragement and liberality, that it is difficult, at present, to distinguish between the work of the artists of their day, and the real antiques. It was especially in cameos that their skill was most frequently exercised, and some of the finest now extant are modern. One of the most famous engravers of the fifteenth century, was Domenico de Milano, usually designated as Domenico de Chamei, who was frequently employed by Lorenzo di Medici. Matteo del Nassaro, another engraver of the following century, also acquired great fame, and was called into France by King Francis I. Cameo ornaments for the head were then fashionable, and the head of Dejanira, engraved in relievo on a very fine agate by that artist, was greatly admired. He made excellent use of the different shades of color the stone presented, to represent in their natural colors, the flesh, the hair, the lion's skin; a red streak in the stone was so skilfully taken advantage of for the inner side of the skin, that it gave it the appearance of having been newly flayed.

The demand for cameos which this fashion occasioned was such, that fine sardonyxes became scarce, and, to supply their place, recourse was had to shells that offered similar layers of different colors. This substance being much softer than agate, was easier wrought, and the price of ornaments of shell-cameos was much inferior to those of stone; they were, however, much inferior also in point of wear, as the least contact with other substances altered their beauty. A very pretty set of this kind, is the necklace that belonged to Diane de Poitiers, and which is now in the Bibliothèque Impériale. It is composed of fourteen small cameos on shell; in the middle is an agate, on which is the portrait of the celebrated beauty represented as the goddess of the chase; the attributes are in diamonds.

Many shell-cameos are still made in Italy, the fine agate-onyx becoming daily more scarce.

Fraudulent restorations are often made by carefully cutting out the graven portion of antique stones, and applying it on the plain ground of an agate of another color, this giving it the appearance of a cameo on an agate-onyx. Another fraud, often successfully practised, is the imitation of the sardonyx by the application of a shell engraved and applied on a hard stone.

Among antiques, the most valued are those which bear the artist's name, and this preference has given rise to spurious signatures being applied to them by dishonest dealers. The trick is not new; it was practised in the time of Phædrus, who complains of it in one of his fables. Much care is required to detect the cheat; the shape of the letters will often betray the origin. Some amateurs of the last two centuries, following in this the example of Lorenzo de Medici, have had their names engraved on their antiques to indicate the ownership; and the celebrated Maffei was at great pains to interpret the letters L. A. V. R. M. E. D. which he had found on several stones that had pertained to that Grand Duke of Tuscany

Among the moderns, Lorenzo de Medici was the first to make a collection of graven stones, and this, subsequently enriched by Cosmo and his successors, is now in the splendid gallery of Florence.

The example given by the Medici has been followed in other parts of Europe. Collections were formed by sovereigns, by wealthy amateurs, savants, and artists. The most famous of the present day, are the Cabinet of the Imperial Library of Paris; the collection in the Gallery ot Florence, which numbers one thousand stones; that of the Vatican in Rome; those of the King of Prussia, the Emperor of Austria, the Council of Leipsic, the King of Denmark, the Castle of Rosenburg at Copenhagen, and that of the Emperor of Russia, which contains the cabinets of Patter and Orleans. Among private collections may be quoted, the ancient collections of Strozzi and Ludovici in Rome, of Poniatowsky in Russia, those of the Dukes of Bedford and Marlborough in England, and those of the Duke of Blacas, of Count Portalès, and of Baron Roger in Paris.

The ancients never engraved on hyaline corindons; engraved gems of this class belong to the modern times. Among the most celebrated is a ruby that belonged to Runjeet Singh, weighing fourteen rupees (half-an-ounce), and on which were engraved the names of several kings, its former possessors; among them were those of Aurungzebe, and Ahmed Shah. Domenico de Chamei engraved on a balass ruby the portrait of Louis the Moor, Duke of Milan.

In Mr. A. S. Hope's collection in the London Exhibition, there was a fine intaglio of Minerva, engraved on an oriental ruby of fifty-three grains. Another engraved oriental ruby of cinque-cento date, in the same collection, bears the head of Jupiter.

When we say that the ancients never engraved on precious stones, we should except the Aztecs. Among the relics of their highly refined civilization, we have seals and rings set with precious stones, on which is engraved the constellation of Pisces.

PART FIFTH.

HISTORICAL JEWELS.

CHAPTER I.

EMINENT PERSONAGES AND THEIR JEWELRY.

Important part played by Jewels in the Lives of the Great.—Solomon's Magic Ring.—Talismanic Ring of Gyges.—Token Ring of Rama.—Ring of Polycrates.—Anecdote of Cæsar's Ring.—Rings sacrificed in Token of Grief.—Jewels presaging Important Events.—Nero's Armlet.—Galba's Necklace.—Galba's Crown.—Rings of Tiberius.—Crowns of Henry III., Louis XVI., and James II.—Jewels celebrated for Size and Beauty.—Three Diamonds of Charles the Bold.—His Jewelled Hat.—His Three Rubies.—Ruby in the Crown of England.—Ruby of Rodolphe II.—Of Elizabeth.—Ruby and Topaz of Runjeet Singh.—Rubies mentioned by De Berquen.—Ruby presented by the Czar to the King of England.—Emeralds of Fernando Cortez.—The Sacro Catino.—Emerald in the Temple of Boudha.—La Peregrina.

> **"Fresh from the merchant, diamonds convey no sentiment but that of wealth: while these hereditary diamonds recall whole generations of stately beauties."—Miss Landon.**

Jewels have played an important part in the lives of the great in all ages. As an omen, as a token, as a symbol, they conveyed joy or sorrow, inspired fear or confidence, and were supposed to give warning of impending events. When the counsels of wisdom and the lessons of experience have failed to move, some little trinket has shaken the firmest nerves, appalled the stoutest courage, melted the sternest heart; in other circumstances, it has strengthened the weak, confirmed the irresolute, emboldened the timid. Rings have been especially associated with the fortunes of man, whatever his station in life. Who has not possessed one of those talismans, within the magic round of which every hope, the happiness, and alas! too often the wretchedness, of a whole life was centered! What pulse has not quickened at sight of this little circlet, preserved, perhaps, through danger, amid exile, and, worst of all, poverty; fraught with memories of past joys, ratifying present happiness, or bringing bright promises of a joyous future! On almost every page of the history of man, we find a jewel connected with the secret springs of his actions, and, whether by accident or design, the arbiter of his destiny. Tiny delegates of those mighty powers,—love and ambition, jewels have overturned thrones and changed the face of empires. When we have diligently traced back to its origin some terrific revolution that, under the specious name of liberty, or the sacred one of religion, has deluged the land in blood, we often find no other great first cause. Unfortunately, in the annals of distant ages, time has done its work, and effaced many a record; still here and there a link of the broken chain may yet be caught, and, comparing the past with the present, we are tempted to exclaim with Montaigne,—"*Les faibelsses humaines ont toujours été les mêmes, elles ne font que changer de nom.*"

Nevertheless we comfort ourselves with the reflection, that we are not more foolish than our forefathers, and that succeeding generations will be no wiser.

* * * *

Among the many fables with which the history of Solomon has been graced by his oriental historians, we have that of his wonderful ring, to the magical properties of which he was indebted for his power over demons and genii. On its *chaton* he beheld whatsoever he desired to know in heaven or upon earth. Having taken off his ring one day when about to enter his bath, it was carried off by a fury, and thrown into the sea. Greatly chagrined by a loss entailing that of his dominion over the supernatural denizens of earth, air, and sea, and deeming himself also deprived of the wisdom that enabled him to rule, the king resolved never to reseat himself on his throne until he had recovered his talisman. At the expiration of forty days, the monarch, who had not, it seems, included fasting in his self-imposed penance, found the precious jewel in the belly of a fish brought to the royal table.

Another mythological jewel was that of Gyges. According to Plato and Cicero, the success of the Lydian shepherd arose from his possession of a magic ring, which he had found in the body of a bronze horse. By turning its *chaton*, the wearer became invisible. Having, by means of his talisman, gained access to the beautiful queen and seduced her affections, he murdered king Candaules, her husband, and reigned thirty years on his usurped throne.

In the Hindoo poem of the *Ramayana*, a ring plays an important part. When Sita is carried off by the demon Rhevan, the only creature who could find out the spot where the captive fair was concealed, was the monkey. Skipping over the sea into the gardens of Rhevan, he communicated to the weeping beauty the message with which he had been entrusted by Rama, her disconsolate husband. But Sita listened incredulously, until he confirmed his words by producing a favorite ring, the gift of Rama, and which she had left behind her. For this and other important services in the same good cause, is the monkey held in honor to this day among the Hindoos.

The ring which Polycrates, the tyrant of Samos, threw into the sea, in the hope that this voluntary sacrifice of that which he held as the most valuable of his possessions would balance his account with the fickle goddess, was, like that of Solomon, brought back by an honest fish. Pliny, who had seen the ring, says it was a sardonyx, and not engraved. It was presented by Livia, the wife of

Augustus, to the temple of Concord in Rome, where it was kept in a golden horn. Many other rings there were more prized than this.

The next in reputation among ancient rings was that of Pyrrhus, who warred with Rome. The stone was an exceedingly beautiful agate, on which a freak of nature had represented as perfectly as though done by man, the nine Muses, with their attributes, and Apollo with his lyre. According to Pliny, there were no other famous rings extant in his day. Ismenias, the flute-player, had several fine gems, and of his vanity and folly the following anecdote is related. Having heard that an emerald, on which was engraved the Danaida Amymone, was for sale in the island of Cyprus, for the sum of six gold denarii, he sent that sum to purchase it. Ere his messenger arrived, the owner had lowered his price and therefore returned him two denarii. Far from feeling gratified, Ismenias exclaimed that the merchant was a fool, and had greatly diminished the merit of the stone.

When Cæsar harangued his troops after his passage of the Rubicon, he often held up the little finger of his left hand, protesting he would willingly pledge even to his ring to satisfy the claims of those who would jealously defend his cause. The soldiers in the farthest ranks, who could see the orator, but were too far to hear his words, mistaking the gesture, imagined he was promising to each man the dignity of Roman Knight, and the sum of five hundred thousand sesterces. How far this error contributed to his success it would be difficult to say, but there is no doubt it had its influence.

Among the tokens of grief manifested at the obsequies of Cæsar, when all classes threw whatsoever they had most valuable about them on his funeral pyre,—when the very seats of the magistrates fed the flames into which the veterans of the old legions cast their cherished weapons,—the Roman matrons made what to them were equally costly sacrifices to his manes,—they burned their personal ornaments, the robes and *even* the rings of their sons.

At the death of Augustus, among many other measures proposed in order to honor his remains, it was suggested that persons entitled to wear a gold ring, should on the occasion of his funeral exchange it for an iron one.

Conscious that his last hour was approaching, the Emperor Tiberius took his ring from his finger and held it some time, as though intending, but hesitating, to give it to some one—an act equivalent to naming his successor; he finally put it on again, and clenching his hand, remained long motionless and insensible. Reviving at last and finding his ring had been taken from him, he demanded it back; upon which his attendants smothered him with the cushions.

Amid the inauspicious presages that announced the fate of Nero, was the gift he received from Sporus of a ring, on which was engraved the rape of Proserpine.

During the childhood of Nero, the skin of a snake was found near his bed, having probably been cast there by the animal. Accepting this incident as a favorable omen, Agrippina caused the reptile's spoil to be enclosed in a golden armlet, which she enjoined her son to wear constantly on his right arm. When the memory of the mother he had destroyed became odious to him, the prince threw aside the sacred jewel. But in the dread hour of retribution, remembering the supernatural virtue supposed to reside in the discarded armlet, he anxiously sought for it; in vain, however, it had disappeared for ever, and with it all hope of salvation.

A necklace is supposed to have occasioned the death of Galba. He had selected from his treasures a necklace enriched with pearls and precious stones, intending it for the statue of Fortune in his country house at Tusculum; on second thoughts, however, he deemed the ornament too rich for a statue in that dwelling, and dedicated it to Venus in the capitol. On the following night, Fortune visited his dreams, and bitterly reproaching him with having deprived her of the promised gift, threatened that she also would resume the favors she had bestowed on an ungrateful favorite. The alarmed Emperor rose with the dawn, and despatched officers to his house at Tusculum to prepare a sacrifice; he himself soon followed, anxious to avert the fulfilment of the sinister presage. On his arrival, to his consternation and surprise, he found but a heap of ashes on the altar; on the steps stood an old man, clothed in black, who, in one hand held a crystal vase containing incense, and in the other a clay vase with wine.

Another omen foreshadowed the death of this Emperor. It was remarked that, while he was offering a sacrifice to Jupiter, on the first day of January, his crown fell from his head. A similar accident occurring in modern times, has always been viewed with like feelings of superstitious awe, as an evil portent, and, by some strange coincidence, has been followed by disastrous results.

Henry III. of France, when the crown was placed on his head, exclaimed:—"*Elle me blesse!*" it wounds me.

Louis XVI., when the diadem touched his brow, hastily raised his hand as though to remove it, saying:—"*Elle me gêne!*" it hurts me.

When James II. was crowned, the diadem fitted so ill, that it required the assistance of one of the attendants, to prevent it from falling off. The circumstance made a deep impression on the minds of those present, and especially on that of the queen.

Many curious historical gems have unfortunately been lost, or at least the circumstances that invest them with a peculiar interest are unknown to their present possessors. Of these are a number of the superb jewels of Charles the Bold. When Louis de Berquen revived, or, as was long thought, invented the art of polishing and cutting the diamond, he was entrusted by the Duke with three fine stones to be dealt with according to his skill. The Duke was so much pleased with the result, that he rewarded the artist with the munificent sum of three thousand ducats. One of these gems the prince presented to Pope Sextus IV.; another, in the shape of a heart held aloft by two hands, and set in a ring, he gave, as a symbol of faith and amity, to his treacherous neighbour, Louis XI.; the third, a thick gem, he wore on his finger the day he was killed, just one year after the gems were cut.

This last diamond was probably the only clue to the recognition of the corpse of the wearer. The body of the great Duke of Burgundy, who never knew fear, or as was usually said of him, who only feared the fall of the heavens, was not found until the 7th of January, three days after the disastrous battle of Nanci. A laundress belonging to the prince's household having joined the sad searchers, recognized the glittering gem on the finger of a corpse, the face of which was buried in the snow. She sprang forward, and turning over the body, exclaimed: — "Alas, my prince!" The ring and other tokens completely identified the Duke, although the face was terribly disfigured; in the endeavour to disengage the head from the ice in which it was prisoned, the skin had stripped off; a fearful gash had cleft the head from ear to mouth on one side, and on the other, the cheek had been partly gnawed by wolves.

The yellow velvet hat, *à l'Italienne*, worn usually by Charles, and lost at the battle of Granson, was surrounded by a coronet of precious stones of inestimable value. This hat was picked up by one of the rude victors, who, after placing it jestingly on his head, threw it from him, saying, he would much rather have had a suit of stout armour to his lot. Jacques Fugger bought it, and sold a large portion of the jewels, some years after, to the Archduke Maximilian, the husband of Charles' daughter, Marie of Burgundy, the natural heiress of all these riches.

Of many other famous jewels of Charles, lost at Granson

or Nanci, the traces have been lost. He had three rubies, called the "Three Brothers," and two called "*La Hotte et la Balle de Flandres.*" In a letter of James I., that sovereign mentions among other jewels which he sends to his son, the "Three Brethren." The similarity of name would seem to identify these as the rubies of the Duke of Burgundy.

A very celebrated ruby is the large heart-shaped balass, set under the back cross in the crown of England. It is in its natural shape, and has received no polish from art. Its color resembles a Morella cherry, dark red, and semi-transparent. This fine gem was brought from Spain by Edward the Black Prince, when he returned from the expedition he had undertaken there, to assist the right of the king Don Pedro of Castile, against his usurping brother. This ruby was subsequently worn by Henry V. at Agincourt.

The Emperor Rodolphe II. possessed an oriental ruby the size of a hen's egg, which he had inherited from his sister, the widow of the King of France. This ruby was considered the largest in Europe.

Sir James Melville, in his *Historic Memoirs*, says, that Queen Elizabeth showed him "a fair ruby, great, like a racket ball."

Runjeet Singh had a large ruby weighing fourteen rupees, with the names of several kings engraved upon it, and among them those of Aurungzebe and Ahmed Shah. He also had a topaz as large as half a billiard-ball, which had cost him twenty thousand rupees.

Robert de Berquen, in his *Merveilles des Indes Orientales et Occidentales*, speaks of several stones of extraordinary size and beauty. "Josephus Barbaro, a Venetian gentleman, says in a report made to the Signori of Venice, that when he was ambassador for the Republic at the court of Yussum Cassan, King of Persia, on a certain day of the year 1472, when he was received in solemn audience, that prince showed him a handkerchief filled with the rarest and most inestimable precious stones. Among others there was a table-cut balass ruby, of a beautiful shape, of at least a finger's breadth, weighing two ounces and a half, and of a most peerless color: in fact, it was a perfect paragon, so exquisite, that when the King asked what he valued it at, he replied, he thought a city, or even a kingdom, would scarcely pay it."

De Berquen goes on to tell of rubies far surpassing even the jewel of the Persian monarch:—

"A person of rank of this city (Paris) has three (rubies), which if the king (of Persia) had possessed them he would have been far more vain of. The one had once been set in a gold crown covered with gems, with which Pope Stephen V., who came to France in 817, crowned the King of France, Louis le Débonnaire, in Rheims, as Emperor—a ceremony not performed in that city since Clovis; and this ruby is in the shape of a lozenge, weighing one hundred and twenty-three carats and a half. The other, which is in the shape of an egg, weighs two hundred and forty-four carats and three-quarters; and was given by the Neapolitans, in 1264, in the time of St. Louis, to Charles, Duke of Anjou, the king's brother, when that prince had hunted Manfred out of Sicily. The third, weighing two hundred and nine carats, had belonged to Anne, Duchess of Brittany, who married Charles VIII., in 1491; and brought with her, among other rings and jewels, this ruby." How these precious rubies had fallen into the hands of a private individual, or what became of them afterwards, De Berquen does not say.

In a letter from Mr. Wanley to Dr. Charlett, it is said, that when the Czar, Peter the Great, left England, he presented the King with "a rough ruby, which the greatest jewellers of Amsterdam, (as well Jews as Christians,) valued at two thousand pounds sterling. 'Tis bored through, and when it is cut and polished, it must be set upon the top of the imperial crown of England." With his characteristic simplicity, the Russian monarch carried this valuable gem to the English monarch in his waistcoat pocket, and presented it wrapped in a piece of brown paper.

Among the spoils brought by Fernando Cortez, from the province called the Golden Castile, were five emeralds, then valued at one hundred thousand crowns, but which at the present day would be priceless. The first was cut in the shape of a rose with its leaves; the second, in that of a hunting horn; the third, in that of a fish with golden eyes; the fourth, was a bell, the clapper of which was a large pear-shaped pearl; around the edge, which was of gold, were these words in Spanish,—"*Blessed be he who made thee;*" the fifth and most precious of all, was a cup on a golden foot, with four little gold chains attached to a large pearl by which the jewel was hung as an ornament to the person. Around the brim of the cup, which was of gold, was engraved the following inscription,—"*Inter natos mulierum non surrexit major.*"

These jewels were perhaps in some measure the cause of the loss of their owner's court favor. It is said that the empress-queen, wife of Charles V., had expressed a desire to possess jewels more remarkable for the extraordinary workmanship displayed in their cutting, than even for their intrinsic value. The conqueror of Mexico was, however, about to be married; the bride was young, fair, and noble, and the lover was unwilling to forego the pleasure of presenting to his betrothed, gems that had excited the admiration and envy of all the court. This preference alienated the royal regard.

The fatal armlets of Rienzi would be precious mementos of the last of the Roman tribunes, had they been preserved to our day. When the conspirators surrounded the capitol, Rienzi attempted to escape, disguised in a peasant's coat, with his face blackened, and a quilt and other bed furniture on his head and shoulders. Chance or treachery betrayed him. A man seeing him on the steps, looked at him earnestly, and then seizing him by the arm, held him fast. Unfortunately the bracelets the senator wore, and which he had forgotten to take off, betrayed him.

A prodigy, held to be somewhat apocryphal in these sceptical times, is the famous Genoese dish, of one entire and perfect emerald, known as the Sacro Catino, out of which *Our Saviour eat the Last Supper!* This rather expensive piece of table gear for the *ménage* of a Jewish publican, had been brought away by the royal crusaders, when they took Cesarea, in Palestine, under Guillaume Embriaco, in the twelfth century. In the division of the spoils, this emerald fell to the share of the Genoese crusaders, into whose holy vocation something of their old trading propensities evidently entered; and they estimated the vulgar value, the profane price, of this sacred relic so high, that, on an emergency, they pledged it for nine thousand five hundred livres. Redeemed and replaced in the church of San Lorenzo, it was guarded by knights of honor called *Clavigeri*, and was exhibited once a year, with great pomp. Millions knelt before it; and the price imposed on him whose ardent zeal touched it with a diamond, was a thousand golden ducats.

The French, having seized this precious treasure, carried it to Paris, where it was sacrilegiously submitted to the examination of mineralogists. Instead of submitting it, with its traditional story, to a Council of Trent, they handed it to the "Institut," and chemists, geologists, and philosophers were called upon to decide the nature of that relic, which bishops and priests had declared to be too sacred for human investigation, or even for human touch. The result of the scientific investigation was, that the *Emerald Dish*, was a PIECE OF GREEN GLASS.

When the Congress of Vienna, in 1815, made the King of Sardinia a present of the dukedom of one of the oldest

Republics in Europe, and restitutions were making *de part et d'autre*, Victor Emmanuel insisted on having his emerald dish; not for the purpose of putting it in a cabinet of curiosities, as they had done at Paris, to serve as a curious memento of the remote epoch in which the art of making colored glass was unknown, but of restoring it to the shrine of San Lorenzo—to its guards of knights servitors, to the homage of the people, with a republished assurance that this is the invaluable *emerald dish*, the "Sacro Catino" *which the Queen of Sheba offered, with other gems, to King Solomon*, (who deposited it in the Temple,) and which afterwards was reserved for a higher destiny still.

The Italians account for the emerald dish having proved a failure in the hands of the French spoilers by saying, that the latter had been completely duped—*mystifié*: the guardians of the precious relic had secreted it, and provided an imitation, which was the glass dish that travelled to Paris.

A modern traveller tells, however, of an emerald which, if it cannot boast of having been consecrated by the touch of divinity, like the "Sacro Catino," is still more remarkable for size. Indeed, the story is so incredible, that it is best told by the author:—"In the temple of Bondha, in Siam, there is the figure of the god two feet high, said to be cut of a single emerald. This idol had two brilliants, flashing light through the temple, in place of eyes, which cost in Brazil twenty thousand dollars. The value of the whole god is inestimable. I doubted its genuineness, but Prince Momfanoi assured me it was an emerald, and not a beryl as I suggested*."

The famous pearl, "la Peregrina," belonging to the crown of Spain, and accounted one of the most beautiful in the world, was brought from Panama, in 1560, by a

* *Narrative of a Voyage Round the World in* 1835-6-7. By W. S. W. Buschenberger, M.D.

cavalier named Don Diego de Temes, who presented it to Philip II. It is about the size of a pigeon's egg, and shaped like a pear. It was then valued at fourteen thousand ducats; but Freco, the king's jeweller, having seen it, said it might be worth £14,000, £30,000, £50,000, £100,000, as such a pearl was priceless. Its size and beauty caused it to be named the *Peregrina*, and it was exhibited in Seville as a phenomenon.

This magnificent pearl is said to have been fished by a little negro. The oyster that contained it was so small, that the finder, thinking it could not contain any pearls, was about to throw it back into the sea unopened; fortunately, he thought better of it, and for the discovery of the treasure it contained, was rewarded with his liberty. In return for this splendid gift, the cavalier, his master, was rewarded by the king with the post of grand provost of Panama.

When Philip V. was endeavouring to maintain his right to the Spanish throne, the queen, obliged to retreat from Madrid, at the approach of the archduke, confided all her jewels to a French valet, to carry into France. Among these valuables was the priceless Peregrina. Vasu, the man in whom such confidence was placed, proved as faithful as the Spánish grandees who so well defended their sovereigns. When the crisis was over, the jewels were brought back to Spain.

When the tomb of Charlemagne at Aix-la-Chapelle was opened, his bones were found enveloped in Roman vestments; his double crown of France and Germany was on his fleshless brow; and his pilgrim's wallet was by his side, as well as his good sword *Joyeuse*, with which, according to the monk of St. Denis, he clove in two a knight clothed in complete armour. His feet rested on the buckler of solid gold given to him by Pope Leo, and round his neck was suspended the talisman which rendered him victorious, and which was formed of a piece of the true cross, sent to him by the Empress Irene. It was enclosed in an emerald, attached to a large chain of gold links. The burghers of Aix-la-Chapelle presented it to Napoleon when he entered that town in 1811. One day, in playful mood, he threw it over the neck of Queen Hortense, declaring that he wore it on his breast at the battles of Austerlitz and Wagram, as Charlemagne had worn it for nine years. From that day, the Duchess of St. Leu never laid aside the precious relic.

CHAPTER II.

THE FINEST DIAMONDS KNOWN.

The Paragon Diamonds.—Diamond of the Rajah of Mattan.—The Orloff, or Grand Russian.—The Grand Tuscan.—The Regent, or Pitt.—The Star of the South.—The Koh-i-noor.—The Shah of Persia.—History of Three Diamonds of Charles the Bold, including the Sancy.—The Nassuck.—The Pigott.—The Blue Diamond.—The Crown Jewels of Spain and Brazil.

There are few diamonds in the world that exceed a hundred carats in weight. Such were formerly termed *paragons*; their number is exceedingly limited, the following being the only ones known.

The diamond of the Rajah of Mattan, in Borneo, the largest in the world, is still uncut, however: it weighs three hundred and sixty-seven carats, or two ounces, 169·87 grains troy.

The Orloff, or Grand Russian . .	193	carats.
The Grand Tuscan	139½	„
The Regent, or Pitt	137	„
The Star of the South	125	„
The Koh-i-noor	102	„

After these, the following stones are celebrated for size and beauty:—

The Shah of Persia	86$\frac{3}{16}$	carats.
The Sancy	53½	„
The Nassuck	89	carats 1½ grs.
The Arcot Brilliants . . .	56	„
The Pigott	49	„

The largest diamond known, that of the Rajah of Mattan, has never been brought to Europe, and few particulars have been related of it. It is shaped like an egg, with an indented hollow near the smaller end. It is said to be of the finest water. Many years ago, the governor of Batavia tried to effect its purchase, and sent Mr. Stewart to the Rajah, offering one hundred and fifty thousand dollars, two large war brigs, with their guns and ammunition, and a considerable quantity of powder and shot. The Rajah refused, however, to despoil his family of so rich an inheritance, to which the Malays, indeed, superstitiously attach the miraculous power of curing all kinds of diseases by means of the water in which the diamond is steeped; and with it, they believe, the fortune of the family is connected.

The *Orloff* probably once belonged to the Great Mogul, and when that prince was conquered by Nadir, the Shah of Persia, this gem, which bore the name of the "Moon of the Mountain," fell, together with other jewels, into the hands of the victor. The Shah having been assassinated in 1747, the crown-jewels were plundered and secretly disposed of.

A man named Shafrass, commonly known at Astracan as the "Man of Millions," then (in 1747) resided at Balsora with his two brothers. One day, a chief of the Anganians applied to him, proposing to sell him privately, for a very moderate sum, the above diamond, together with a very large emerald, a ruby of very considerable size, and other precious stones of less value. After some hesitation, the Armenian concluded the bargain by paying fifty thousand piasters for all the jewels.

At first the brothers were afraid to dispose of their acquisitions; but, after a period of ten years had elapsed, the eldest proceeded to Amsterdam, and there publicly offered the jewels for sale.

The English and the Russian governments were the highest bidders for the large diamond, and it was at last purchased by the Count Orloff, for the Empress Catherine, for four hundred and fifty thousand roubles, ready money, and a grant of Russian nobility.

According to another account, this beautiful gem was one of the eyes of the idol Scheringham, in the temple of Brahma. The fame of these bright eyes having reached the ears of a French grenadier of the garrison of Pondicherry, he deserted, took refuge with the Brahmins, and having adopted their religion, and conformed to their rites, finally succeeded in possessing himself of one of the coveted jewels. Escaping to Madras, he sold his booty to a sea-captain for fifty thousand francs; the captain disposed of it to a Jew for three hundred thousand francs; and the Jew to the Armenian for a much larger sum. From the hands of Shafraes, it passed into the possession of the Empress Catherine in 1772, on the terms stated above. This jewel is by some called the Lazareff diamond; and another version of its history supposes it to have been brought to Russia by Lazarus Lazareff, the head of the Armenian family of that name, and grandson of Manouk Lazareff, treasurer of Shah Abbas II. On the sum paid, and the patent of nobility granted by Catherine, in exchange for the gem, both accounts agree.

The size of this gem is about that of a pigeon's egg; its lustre and water are very fine, but its shape is defective.

The *Austrian Diamond*, also called the *Maximilian* and the *Grand Tuscan*, became by purchase the property of the Grand Dukes of Tuscany. It was long preserved in the family of the Medici, but at last passed into the hands of the Emperor whose name it bears, since which time it has been preserved as a heir-loom in the imperial family. It is cut as a rose, is nine-sided, and covered with facets, presenting a star with nine rays. It has a yellowish tint that greatly lessens its value. It weighs one hundred and thirty-nine and a half carats, and is valued at £155,682.

The *Regent*, although not the largest, is looked upon as the most perfect and beautiful diamond in Europe, being remarkable for its shape, proportions, and fine water. In its primitive shape, it weighed four hundred and ten carats; the cutting of it as a brilliant, which cost two years' labor, was performed at an expense of three thousand pounds, and reduced its size to one hundred and thirty-seven carats. This diamond, which also bears the name of the *Pitt*, from its first European owner, was stolen from the mines of Golconda, and sold to Thomas Pitt, grandfather of the Earl of Chatham, when that gentleman was governor of Fort St. George, in the East Indies. After having offered his diamond to several sovereigns, the difficulty of finding a purchaser induced the owner to lower his demands, and the Duke of Orleans, then Regent, finally consented, at the solicitation of the famous Law, to purchase it for Louis XV. The price was two million three hundred thousand crowns—(ninety-two thousand pounds); the owner reserving the fragments taken off in the cutting, and which made several fine diamonds, worth several thousand pounds sterling. Five thousand pounds were spent in the negotiation. This diamond is estimated at twice the amount paid for it.

A rumour had been circulated in England, that the Governor of Fort St. George had not come fairly by this jewel, and it was alluded to by Pope, in the lines:

> "Asleep and naked as an Indian lay,
> An honest factor stole the gem away."

In refutation of this calumny, Mr. Pitt published a statement of the transaction by which the gem came into his possession, for the sum of twenty thousand pounds.

When, after the fall of the throne of Louis XVI., the people, or rather the populace, insisted that the beautiful works of art and nature hitherto reserved to the enjoyment of the refined and the educated, should be exposed to the gaze of the meanest and least capable of appreciating their merits, the Regent was paraded to the mob. So little did the exhibitors confide in the integrity of these patriots, however, that great precautions were taken to prevent the consequences of too strong an attraction. The passer-by who chose to demand, in the name of the sovereign-people, a sight of the finest of the ex-tyrant's jewels, entered a small room, within which, through a little wicket, the *national* diamond was presented to the admiration of the citoyen or citoyenne in tatters; a strong steel clasp fastened the profaned paragon to an iron chain, the other end of which was secured within the aperture through which it was handed to the patriots. Two policemen, disguised as gendarmes, kept a vigilant watch on the momentary possessor of the gem, until, having held in his hand the value of twelve million francs, according to the estimate in the inventory of the crown-jewels, he again took up his hook and basket at the door, and resumed his scrutiny of the dirt-heap at the street corner.

The Regent, pawned by Napoleon I., stolen by a band of robbers, made by Talleyrand a bait to seduce Prussia, passing unscathed through half-a-dozen revolutions, still pertains to France. The first Emperor wore it mounted in the hilt of his state sword; it is now set in the imperial diadem.

The *Star of the South*, the largest diamond as yet brought from Brazil, is in the possession of the king of Portugal. It weighed uncut two hundred and fifty-four and a half carats. The grandfather of the present king of Portugal, who had a passion for precious stones, had a hole bored through it, and wore it rough, suspended about his neck, on gala days. It was estimated by Romé de l'Isle at the enormous sum of three millions sterling! It now weighs one hundred and twenty-five carats, and is of an octahedron form.

The history of the discovery of this stone is romantic. Three Brazilians, Antonio de Souza, José Felix Gomez, and Thomas de Souza, were, for some imputed misdemeanor, condemned to perpetual banishment into the wildest part of the interior. The region of their exile was the richest in the world; every river rolled over a bed of gold, every valley contained inexhaustible mines of diamonds. The unfortunate men cheered the horrors of their exile by the hope of discovering some rich mine that would obtain the revocation of their condemnation. For six years they continued their search, and fortune was at last propitious. An excessive drought had lain dry the bed of the river Abaiti, about ninety-two leagues south-west of Serro do Frio, and here, while working for gold, they discovered a diamond of nearly an ounce in weight. Overwhelmed with joy at this god-send, they resolved to proceed at all risks to Villa Rica, and trust to the mercy of the crown. The governor, on beholding the magnitude and lustre of the gem, immediately appointed a committee of the officers of the diamond-district to report on its nature; and on their pronouncing it a real diamond, it was forthwith despatched to Lisbon. It is needless to add, that the sentence of the finders was revoked.

The *Koh-i-noor*, now weighing one hundred and two carats, and consequently the sixth in size of the paragon diamonds, is supposed to have once been the largest ever known, and the same seen by Tavernier among the jewels of the Great Mogul. Its primitive weight was nine hundred carats. It was unfortunately put into the hands of Hortensio Borgis, a Venetian diamond-cutter, who wasted the precious substance so inconsiderately, that, although he made no attempt at brilliant-cutting, and merely surface-cut it, he reduced it to two hundred and eighty carats. The enraged monarch, in lieu of paying him for his labor, fined him ten thousand rupees, and would have made him pay still dearer, had the unfortunate jeweller possessed wherewith. Tavernier adds, that had the Venetian been skilful, he might have reserved good cuttings for himself, without doing the king any injury, have spared himself much labor, and left the stone much larger.

Though the descent of the Koh-i-noor has not been very satisfactorily traced down to the present day, it is confidently asserted that this famous gem belonged to Karna, king of Anga, *three thousand and one years ago!* No one would presume to doubt the accuracy of a statement so minute, that it reckons even the odd year in its computation. At any rate, it would be difficult to prove the contrary.

According to Tavernier, this gem was presented to Cha-Gehan, the father of Aurungzebe, by Mirzimola, when that Indian general, having betrayed his master, the king of Golconda, took refuge at the court of the Great Mogul. Since it was admired by the French traveller, this diamond has passed through the hands of several Indian princes, and always by violence or fraud. The last Eastern possessor was the famous Runjeet Singh, king of Lahore and Cashmere, from whom it passed into the hands of the English on the annexation of the Punjaub: it was brought to London in 1850.

The king of Lahore had obtained this jewel in the following manner: having heard that the king of Cabul possessed a diamond that had belonged to the Great Mogul, the largest and purest known, he invited the fortunate owner to his court, and there, having him in his power, demanded his diamond. The guest, however, had provided himself, against such a contingency, with a perfect imitation of the coveted jewel. After some show of resistance, he reluctantly acceded to the wishes of his powerful host. The delight of Runjeet was extreme, but of short duration, the lapidary to whom he gave orders to mount his new acquisition, pronouncing it to be merely a bit of crystal. The mortification and rage of the despot were unbounded; he immediately caused the palace of the king of Cabul to be invested, and ransacked from top to bottom. But, for a long while, all search was vain: at last, a slave betrayed the secret; the diamond was found concealed beneath a heap of ashes. Runjeet Singh had it set in an armlet, between two diamonds each the size of a sparrow-egg.

Since the Koh-i-noor came into the possession of the English, it has been re-cut, an operation that has greatly improved its brilliancy and general appearance, at the expense of more than a third of its weight. The flaws and yellow tinge that marred its beauty have been removed, and the form of a brilliant given to it, thus bringing out all its lustre. The re-cutting was commenced on July 16th, 1852, the late Duke of Wellington being the first person to place it on the mill. The operation was finished on September 7th, having taken thirty-eight days to cut, working twelve hours per day, without cessation.

Mr. Tennant is of opinion that this celebrated stone, even in the state in which it was first brought to England, was only a portion of the original diamond of that name. In confirmation of his assertion, he quotes Dr. Beke:—"At the capture of Coochan, there was found among the jewels of the harem of Reeza Kooli Khan, the chief of that place, a large diamond slab, supposed to have been cut from one side of the Koh-i-noor, the great Indian diamond in the possession of Her Majesty. It weighed about one hundred and thirty carats, showed the marks of cutting on the flat and largest side, and appeared to correspond with the Koh-i-noor." According to Mr. Tennant, the great Russian diamond singularly corresponds with the Koh-i-noor, and it is not improbable that they all formed one crystal; and that, when united, they would, allowing for the detaching of several smaller pieces in the process of cleaving, make up the weight described by Tavernier.

The *Shah of Persia*, which takes its name from its having been presented by the Persian monarch to the

Emperor Nicholas, is remarkable for the beauty of its water and its lustre, but the shape is singular, being that of a long prism, and its cutting is very irregular. It weighs eighty-six carats and three-sixteenths, and is valued at two hundred and twenty thousand francs.

The chief interest attached to this gem lies in the inscriptions which it bears of its former owners, engraved as follows:—

Ek-Bek Schak, Nizim Schak, Feth Ali Schak,	Lords of Irostan.

There is, perhaps, no gem of the history of which so many different versions have been given as the *Sancy*; the only way, in fact, in which these *variantes* can be accounted for is by the supposition that there have been several gems, of nearly the same weight and shape, to which the same name has been given. There seems, however, to be but one extant at present, and even that appears and disappears from time to time in the most unaccountable manner. Of this diamond, which has been extant for four centuries, very romantic circumstances are related.

The first European possessor of the Sancy was the last Duke of Burgundy, Charles the Bold, who usually wore it, mounted between three balass rubies, on his neck. At that time, the Duke possessed three of the finest diamonds in Europe, and of these the Sancy was the smallest. The history of these three great diamonds is, as M. de Barante very justly remarks, well worth recording; the celebrity of the gems themselves, and the pride since attached to their possession, sufficiently testify the splendour of the princes of the House of Burgundy, for whose spoils the greatest monarchs contended.

After the disastrous defeat of Granson, the Duke's great diamond, supposed not to have its equal, and which had once adorned the crown of the Great Mogul, was picked up on the road, where it had probably been dropped by one of the Duke's fugitive attendants. It was enclosed in a small box adorned with pearls. The finder, a common soldier, kept the box and threw away the diamond as a mere bit of glass. Changing his mind, however, he went back, and finding it under a waggon, sold it for a crown to the curate of Montagny. By the curate, the gem was sold to a citizen of Berne for three crowns. Subsequently another citizen of Berne, of the name of Bartholomew May, a rich merchant, who traded with Italy, offered to William Driesbach a present of four hundred ducats, for that Driesbach had negotiated for him the purchase of this diamond for five thousand ducats. In 1482, the Genoese purchased the stone for seven thousand ducats, and sold it for double that sum to Louis Sforza the Moor, Duke of Milan. After the fall of the house of Sforza, this gem passed into the possession of Pope Julius II. for the sum of twenty thousand ducats. It adorns the tiara: its size is about that of half a walnut.

The next diamond in size was nearly equal to the first, and was purchased by the famous merchant, Jacques Fugger, who kept it some time. Soliman Pacha and Charles V. both bargained for it; but Fugger did not choose to have the gem lost to Christendom; neither did he like to sell it to the Emperor, who was already deeply in his debt. Finally, the diamond was purchased by Henry VIII.; through his daughter Mary, it passed into the possession of Spain, and thence returned to the House of Burgundy into the hands of the great grandson of Charles the Bold. It now belongs to the house of Austria.

The third diamond is the *Sancy*. Of this gem the most contradictory stories are told. The French *Encyclopedia*, published in 1823, says that after the death of Charles, this jewel fell into the hands of the Fuggers, who sold it to Henry VIII., whose daughter Queen Mary brought it in her dower to Philip II., and that since it has not been heard of*! It was not very likely that a diamond of such value could be thus lost sight of; and if, as other writers assert, the Sancy was worn by Louis XIV., it might, according to this last account, have been brought into France, when the Spanish princess, Maria Theresa, married Louis XIV. But in that case there is no origin for the name of Le Sancy, which the diamond derived from its former owner, the Sieur de Sancy, in the reign of Henry IV.

According to another version, James II. possessed it in 1668, when he fled to France.

The most reliable account is that given by Monsieur de Barante, who says, that the Sancy was sold at Lucerne for five thousand ducats, in 1492, and passed into Portugal. When that kingdom was in the hands of the Spaniards, Don Antonio, Prince of Crato, the king of Portugal *in partibus*, visited England and France, in the hope of assistance in prosecuting his claim to the throne. While in England, he pledged a very valuable diamond (the Sancy), which he had brought away with him, to Queen Elizabeth, for five

* It is not probable that the diamond given by Mary to Philip, whether the Sancy or any other, belonged to the English crown-jewels: it must have been her own personal property, for, among the conditions stipulated by Parliament when it consented to the Spanish marriage, the 5th was, that he should carry none of the jewels of the realm out of the kingdom, nor cause any ships or ordnance to be moved. Philip had sent from Spain a very remarkable diamond to his English bride, and this, with other of his gifts, Mary took care to leave him in her will:—"To keep for a memory, one jewel being a table diamond, which the Emperor's Majesty, his and my most honorable father, sent unto me by Count D'Egmont at the insurance (betrothal) of my said lord and husband: also one other table diamond, which his Majesty sent unto me by the Marquis de los Naves, and the collar of gold set with nine diamonds, the which his Majesty gave me the Epiphany after our marriage; also the ruby set in a gold ring which his Highness sent me by the Count de Feria."

thousand pounds sterling. To get rid of the daily importunity of the Portuguese Pretender, who had found the sum insufficient for his purpose, and unwilling to grant him further aid, she was fain to give him back his pledge without repayment of the loan. Nicholas de Harlay, Sieur de Sancy, then (1594), ambassador from Henry IV., to the court of England, either bought the diamond there of Don Antonio, or subsequently in Paris; at any rate, he became the next possessor of it for the sum of seventy thousand francs From this owner, the diamond took its name of Le Sancy; and here the following story, if true at all, takes its place.

When Henry IV. was conquering his kingdom from his subjects, he was, on one occasion, very much in want of money—no unusual circumstance with him. In order to aid his king, a faithful adherent, the Sieur de Sancy, bethought himself of pledging his diamond to the Jews of Metz, and despatched a trusty servant to bring it from Paris; the mission was fraught with peril, the environs of Paris being beset with brigands during the civil wars. The understanding, therefore, between De Harlay and his messenger was, that in case he fell into the hands of thieves he would swallow the diamond. Finding his emissary did not return, the Sieur de Sancy set out to seek him, and having ascertained that his body had been found and buried by peasants, he had it opened and recovered the gem.

The Sancy, after remaining more than a century in the Harlay family, was finally sold to the Regent, and Louis XV. wore it at his coronation. It disappeared from among the crown-jewels in 1789. From that time, there is no trace of the Sancy until 1830, when it reappeared in France, in the hands of a merchant, whose name, and the manner in which he became possessed of the gem have remained a mystery. In 1832, the Sancy was the subject of a law-suit between Count Demidoff and M. Levrat, the gérant and administrator of the Company of Mines and

Foundries of the Grisons in Switzerland. M. Levrat had purchased the Sancy of the Count for six hundred thousand francs, payable in three half-yearly instalments. When the first payment became due, there were no funds, and the Count demanded the cancellation of the sale, and the restitution of the gem, which M. Levrat had deposited at the Mont de Piété. The court ordered the repayment by M. Levrat of the money advanced on the jewel, and its restitution to the Count.

When the suit was before the tribunal, the history of the diamond was given with the most absurd mistakes. The Sancy was said to have belonged to Charles the Bold; then to Antonio, the Portuguese prince, in 1389; then to have been brought from Constantinople by the Ambassador du Harlay, in whose family it remained a century; *at the end of which it was pledged to relieve the necessities of Henry III!* Here came in the story of the faithful servant. Finally it was sold to James II., during his stay at St. Germain's, and by him ceded to Louis XIV.

As Robert de Berquen also mentions having seen the Sancy in England in 1664, it is evident he could not have meant the one so called now, as he describes it as weighing one hundred carats. He says:—"The present Queen of England (Elizabeth) has that (the diamond) which M. de Sancy brought home from his embassy in the Levant, which stone is almost almond-shaped, with facets on both sides, perfectly white and clear, and weighs one hundred carats." The most probable explanation is, that there were two diamonds in the Harlay family; one, the present Sancy—purchased by Nicholas de Harlay of the Prince of Portugal; the other, a larger gem, brought subsequently from Constantinople by his son, Achille de Harlay, when he returned from his embassy in 1617. Both gems bore the surname of the family.

The Sancy which was among the crown-jewels of France, weighed fifty-five carats. If the Sancy which was weighed in Paris in 1836 is the same gem, its bulk is unaccountably diminished, as it now weighs but fifty-three and-a-half carats. It is pear-shaped, and of the finest water. In the inventory of the crown-jewels, made in 1791, the Sancy is estimated at one million francs. It was mounted as a pin.

The *Nassuck Diamond* was among the spoils which were captured by the combined armies, under the command of the Marquis of Hastings, in the British conquest of India, and formed part of what was termed the "Deccan Booty," from its being taken in that part of India which is designated the Deccan. This magnificent diamond is as large as a good-sized walnut, and weighs over eighty-nine carats. It is of dazzling whiteness, and as pure as a drop of dew, but its form is bad. It was sold by auction in 1837, by order of the trustees of the Deccan Prize Money. It is valued at thirty thousand pounds, but was sold to a jeweller for seven thousand two hundred pounds.

The *Pigott Diamond* is of a beautiful shape, cut as a brilliant, weighs forty-nine carats, and is valued at forty thousand pounds. About forty years ago, it was disposed of by lottery, and became the property of a young man, who sold it at a very low price. It was afterwards purchased by the Pacha of Egypt for thirty thousand pounds.

The famous triangular *Blue Diamond*, weighing sixty-seven carats and two-sixteenths, and uniting the most beautiful hue of the sapphire to the most splendid lustre, disappeared from among the French crown-jewels, where it had long held a distinguished place, at the time of the great robbery of those jewels. Notwithstanding the difficulty of disposing of such a stone, it has never been since heard of. From its uncommon beauty and rarity, it was estimated in the inventory at three million francs.

The crown of Spain has long been one of the richest in diamonds. It has not lost much of this wealth, if we may judge by the description of the Queen's appearance at the reception of the ambassadors of Morocco. On this occasion, Her Majesty, Isabel II., is said to have worn diamonds to the amount of *ten million francs.*

The diamonds of the imperial crown of Brazil are, beyond doubt, the most splendid of any crown-possession either in ancient or modern times.

CHAPTER III.

DIFFERENT JEWELS WORN IN ANCIENT AND MODERN TIMES.

Nose, Chin, Cheek, Lip, and Ear Jewels.—Pistols and Poniards.—Jewels of a Daughter of the House of Alba.—Daggers of Eastern Princesses.—Military Collars and Chains of the Romans, Gauls, Mediæval Knights, and of Modern Orders.—Necklace of Penelope; of Eriphyle; of Agnes Sorel; of the Queen of Scots; of the Duchess of Berry; of the present Queen of Prussia; of the Empress Eugénie.—The Girdle an Insignia of Knighthood.—Charmed Girdle of Pedro of Castile.—Diamond Girdle of Isabel II.—Crowns.—Floral Coronals.—Gold Crowns.—Crowns Military Rewards.—Different kinds of Crowns.—Snuff-boxes.—Shoe-buckles.

> " Fair golden tresses grace the comely brain,
> And every warrior wears a golden chain."
> VIRGIL, *Æn.* B. viii. 659.

> " On her white breast a sparkling cross she wore,
> Which Jews might kiss, and infidels adore."
> *Rape of the Lock.*

To enumerate all the jewels with which human vanity has sought to enhance the charms of nature, is almost impossible. Many of the ornaments worn by various nations of antiquity have maintained their place down to the present day, and will probably be worn by future generations to the end of time. Among these are rings, ear-rings, bracelets, chains, necklaces, brooches, clasps, diadems, girdles, hair-pins, aigrettes, &c.

The origin of bracelets and chains, as given by Tertullian, is very poetical, as he attributes the discovery of the material, and the invention of the ornaments themselves, to the love of the fallen angels for the daughters of men.

The choice of jewels has not always been dictated by

good taste. Some oriental nations have delighted in disfiguring themselves in the most barbarous and inconvenient, as well as ludicrous, manner. Among the absurd contrivances invented by folly and sanctioned by custom, are nose, cheek, chin, and lip jewels, worn not only by untutored savages, but by nations that could justly claim to have reached a high degree of civilization. The Peruvians and Mexicans not only wore ear-rings so enormous that the lobe of the ear was frightfully distended, but also lip-jewels, generally of amber set in gold. Tavernier tells us that the ladies of Bagdad wore a *collar* of jewels round the face, and also nose-jewels. The Arab females insert in the cartilage of the nose a ring, of such dimensions that it encircles the mouth and proves no obstacle to the food, as that passes through the ring. This ornament is of the thickness of a goose-quill, but is made hollow to give it the appearance of bulk without its inconvenient weight. The Indian courtesans pierce the left nostril and wear in it a ring set with some precious stone. In the kingdoms of Lars and Ormuz, the females go beyond all this. They pierce the upper portion of the nose, the bone itself, and through this aperture they pass a hook that fastens a sheet of gold shaped to cover the nose, and enriched with emeralds, rubies, and turkoises.

In all the Mahommedan courts, the princes and princesses of the blood royal possess, as a prerogative of their rank, the privilege of wearing two poniards at their belt. Madame de Villars, the wife of the French ambassador at the court of Charles II. of Spain, tells us in her letters that the daughter of the Duke of Alba wore a quantity of jewels, and also *a pistol* fastened to her side by a great knot of ribbons. This young lady was one of the Spanish queen's maids of honor. Madame de Villars does not say whether this warlike ornament was a distinctive badge worn by the nine or ten ladies who held the same post, or was merely a fancy of this noble maiden. "Do not imagine," she says, "that this pistol was a mere trinket; it was half-a-foot long, well-mounted, of finely-polished steel, and quite fitted to do execution."

When the victories of Edward III. had despoiled the French ladies of their jewels, either carried off by the devastating enemy, or sold to redeem their captive lords, their English sisters, decked in these trophies, seemed to imagine they had won them by their own prowess. They assumed conquering airs, and wore diminutive poniards as corsage ornaments, and ugly lawn caps cut like helmets.

Setting aside the trinkets with which folly has occasionally outraged sense, we will briefly examine the origin of those which ancients and moderns have alike admitted to be really ornamental.

The collar, chain, and necklace, are of the highest antiquity, and were worn by Medes, Babylonians, Egyptians, Hebrews, Greeks, and Romans.

Gold collars were given by the ancient Romans to their auxiliary troops, and to strangers, as military rewards; but never any other than silver ones to citizens. This distinction subsequently disappeared, and the metal was in accordance with the rank or deeds of the recipient.

The collars of the Roman knights sometimes fell low on the breast; another kind only encircled the throat.

In a spirit of bravado, the Gauls threw off their clothes before going into battle, and posted themselves naked in the foremost rank, shield and spear in hand*. There was something terrible in the looks of those giants, not one of whom but was tricked out in chains, collars, and bracelets of gold.

* Their descendants, the Zouaves, acted much in the same spirit of daring, on different occasions during the last war with Austria. Taking off their turbans and trowsers, and, leaving them in heaps, they rushed on the enemy in a state of semi-nudity. These *gamins de Paris* have degenerated in size, but not in valour, from their ancestors.

Among the Gauls, as among the Romans, the golden collar was an insignia of knighthood. *Markhok, markhek,* signified knight in Gaelic. The knight was also called *aour-torkhoc,* decorated with the golden collar. Collars are still used as badges of knighthood during the present day; the chains, to which are hung the different orders, are called the *collars* of the orders; there is the collar of the Holy Ghost, of St. Michael, of the Toison, &c.

The chain was, in the East, a badge of honor, and an insignia of authority bestowed by the king himself. Joseph is invested by Pharaoh, and Daniel by Belshazzar, with this ornament. Among the Persians, who had this custom from the Chaldeans, no man dared wear a chain unless in office, and invested by the king.

We are told by Keating, that in the reign of Muirheanhoin, king of Ireland, anno mundi 3070, the Irish gentlemen wore, by royal command, a chain of gold about their neck, to distinguish them from the common people.

The coffin of Edward the Confessor having been opened in the reign of James II., there was found, under one of the shoulder-bones, a crucifix of fine gold, richly enamelled, suspended to a gold chain twenty-four inches long, fastened by a locket of massive gold, adorned with four large stones.

In the middle ages, and till within the last two hundred years, heavy gold chains were universally worn by nobles and gentlemen.

The stewards in great men's houses, and, indeed, many of their retainers, wore gold chains as badges of authority. In the old ballad of *King John and the Abbot of Canterbury,* they are mentioned as worn by the followers of the great churchman to add to his pomp and splendour:

> "A hundred men the king did hear say
> The abbot kept in his house every day,
> And fifty gold chains without any doubt,
> In velvet coates, waited the abbot about."

That chains were very much worn in Shakspeare's time, would be apparent, had we no other authority than his frequent allusions to this ornament. In the *Comedy of Errors* there is a great ado about a chain, and in *Twelfth Night*, Sir Toby bids the steward Malvolio—

> "Go, sir, rub your chain with crumbs."

In *Every Man out of his Humour*, Carlo counselling Soligardo how to appear as a court gallant, tells him he must have a fellow with a great chain. (though it be copper,) to bring him letters, messages, &c. Ben Jonson tells us a beau was not completely apparelled unless he wore a Savoy chain.

Among the jewels sent by James I. to the Prince of Wales and the Duke of Buckingham, in Spain, was a "chayne of gould of eight-and-forty pieces, whereof twenty-four are great and twenty-four small, garnished with dyamonds; and a great George of gould hanging thereat, garnished with dyamonds of sundry sortes; also, one faire chayne of gold, having three-score pieces, with four dyamonds in each piece, and three-score great round pearles."

The gold chain among the Gauls was the chief ornament of those in authority, and in battle distinguished the chiefs from the common soldiers. In the middle ages they were granted as rewards and tokens of sovereign favor; in the nineteenth century, they are the decoration of lord mayors, court-ushers, and parish beadles.

As feminine ornaments we find the chain and necklace very ancient. Homer describes the amber and gold necklace presented to Penelope by one of her suitors; and the fatal gold necklace, set with precious stones, presented by Polynices to the wife of Amphiaraus, to induce her to betray her husband's retreat, is another proof of the favor in which the Greek belles held this ornament. Pliny and St. Clement speak of chains as a feminine ornament.

The wealthy Roman ladies wore them of gold or silver; women of lower rank, of copper. They wound them round their waists as well as round their throats, and hung upon them pearls and diminutive trinkets of all kinds, somewhat as the ladies do at the present day.

In France, necklaces were not worn by ladies until the reign of Charles VII. That prince presented one of precious stones—some say of *diamonds*—to his fair mistress, Agnes Sorel. The gems were probably uncut, perhaps unskilfully set, for the lady complained that they hurt her neck; and, comparing it to an instrument of punishment, she denominated the ornament her *carcan, i.e.*, carcanet. However, as the king admired it, she continued to wear the jewel, saying, that one might surely bear some little inconvenience to please those we love. The fashion was immediately adopted by the ladies of the court, and soon became general.

From that time, the necklace has been more or less worn. Sometimes, as in the reign of Catherine de Medici, pearls were all the fashion; and the pictures of that queen, of the celebrated Diane de Poitiers, her rival, and of the fair Mary Stuart, show how *recherchées* were those ladies in this respect. Under Marie de Medici, pearls continued in favor, not only for necklaces, but every other ornament; dresses were covered with them, and fillets and strings of pearls were mingled with the tresses left to flow loose on the shoulders.

Under Louis XIV., diamonds superseded pearls, and were used with like profusion. Diamond *rivières* took the place of strings of pearls.

The satin-stone necklace, so much worn at the Restoration, at a time when diamonds were beyond the means of the highest in the land, was made and brought into fashion in England. Louis XVIII., when he returned to France, brought a number of these delicate and beautiful ornaments of polished fluor-spar, which he presented to the ladies of his court. The Duchess de Berry, at the time of her marriage, purchased various ornaments of this material to a considerable amount. On the fatal night of her husband's assassination, the duchess had on a satin-stone necklace; and it has been said, with what degree of truth we cannot vouch, that on the anniversary of her widowhood, she never fails to put on a satin-stone necklace, as a memorial of her bridal and subsequent sad loss.

The pearl necklace of the present Queen of Prussia, if preserved unbroken, will be held by future generations as a very interesting memento of the conjugal affection of Frederic William IV. The first year of his marriage, the king presented to his wife, on her birthday, a magnificent pearl; and every succeeding birthday, he added another pearl, of equal size and beauty. After a few years, the pearls formed a collar fitting closely round the throat; at the present day, this superb set of pearls encircles the bosom and falls down to the waist; a very few years more will make the string long enough for a double row.

A former queen of Prussia was not only less fortunate in the love of a husband, but was also deprived of the pleasure, so dear to feminine hearts, of wearing her own jewels. The wife of William I. possessed a quantity of beautiful diamonds, and was extremely fond of displaying them; but this, the absurd tyranny of her husband seldom permitted of her doing. On one occasion, however, he was absent, and his consort eagerly embraced the opportunity, and made her appearance magnificently adorned. There was *grande cour*, and the queen was at cards, when the return of the king was announced. A thunderbolt fallen at her feet would scarcely have caused more alarm. His Majesty had not been expected for four-and-twenty hours. The poor queen, trembling and agitated, without rising from her seat lest she should lose a moment, snatched off all her splendid trinkets and huddled them into her pockets.

When the sovereign of France marries, by virtue of an ancient custom kept up to the present day, the bride is presented by the city of Paris with a valuable gift. Another is also offered at the birth of the first-born.

In 1853, when the choice of His Majesty Napoleon III. raised the Empress Eugénie to the throne, the city of Paris, represented by the Municipal Commission, voted the sum of six hundred thousand francs for the purchase of a diamond necklace to be presented to Her Majesty.

The news caused quite a sensation among the jewellers. Each was eager to contribute his finest gems to form the Empress' necklace,—a necklace which was to make its appearance under auspices as favorable as those of the famous *Queen's Necklace* had been unpropitious. But, on the 28th of January, two days after the vote of the municipal commission, all this zeal was disappointed; the young Empress having expressed a wish that the six hundred thousand francs should be used for the foundation of an educational institution for poor young girls of the Faubourg St. Antoine.

The wish has been realized, and, thanks to the beneficent fairy in whose compassionate heart it had its origin, the diamond necklace has been metamorphosed into an elegant edifice, with charming gardens. Here a hundred and fifty young girls, at first, but now as many as four hundred, have been placed, and receive, under the management of those angels of charity called the *Sisters of S. Vincent de Paul*, an excellent education proportioned to their station, and fitting them to be useful members of society.

The solemn opening of the Maison-Eugénie-Napoléon took place on the 1st of January, 1857.

M. Viron, the *journaliste*, now deputy of the Seine, has given, in the *Moniteur*, a very circumstantial account of this establishment. From it we borrow the following:—

"The girls admitted are usually wretchedly clad; on their entrance, they receive a full suit of clothes. Almost all are pale, thin, weak children, to whom melancholy and suffering has imparted an old and careworn expression. But, thanks to cleanliness, to wholesome and sufficient food, to a calm and well-regulated life, to the pure, healthy air they breathe, the natural hues and the joyousness of youth soon reanimate the little faces; and with lithe, invigorated limbs, and happy hearts, these young creatures join merrily in the games of their new companions. They have entered the institution old; they will leave it young."

The Empress Eugénie delights in visiting the institution of the Faubourg St. Antoine. This is natural. Her Majesty cannot but feel pleasure in the contemplation of all she has accomplished, by sacrificing a magnificent but idle ornament to the welfare of so many beings rescued from misery and ignorance. These four hundred young girls will be so many animated, happy, and grateful jewels, constituting for Her Majesty in the present, and for her memory in the future, an ever new set of jewels, an immortal ornament, a truly celestial talisman.

A fresco painting represents in a hemicycle, the Empress in her bridal dress, offering to the Virgin a diamond necklace; young girls are kneeling around her in prayer; admiration and fervent faith are depicted on their brows.

If there is room left for a wish, it is that the *Maison-Eugénie-Napoléon* may change its name, and be known under that of the *Asyle du Collier de l'Impératrice.*

The fashion of wearing a cross of gold, or set with precious stones, may be traced back to the beginning of the sixteenth century. A portrait of Ann of Cleves shows her adorned with three necklaces, to one of which is attached a jewelled cross. The mode was revived in the beginning of the eighteenth century. The ladies who then went, even to church, in dresses cut very low, wore, as a throat or bosom ornament, small diamond *Saint-Esprits* and crosses. Against this profanation of symbols, a zealous preacher thus indignantly exclaimed from the pulpit:—"Alas! can the cross, which represents the mortification of the flesh, and the Holy Ghost, author of all good thoughts, be more unsuitably placed!" But, like all things preached against, the crosses maintained their place, the ladies now and then adding a heart of the same hard and brilliant substance—not typical of their own, it is supposed.

Clasps were at first only worn by the military, and were used to fasten the mantle; the fashion, however, became general in the third and fourth centuries, when the toga was no longer worn.

Clasps of gold, set with gems, were worn by both sexes; and M. Brutus, in letters written on the plains of Philippi, expressed great indignation at the golden clasps worn by the tribunes.

These clasps were finally made exceedingly large, and loaded with ornaments, something in the style in which brooches and corsage ornaments have been made in modern times.

Roman women wore gold chains, collars, necklaces, bracelets, ear-rings, rings, diadems, fillets, clasps, hair-pins, and ankle-bands. These last, of gold or silver set with precious stones, were only worn by women of an inferior class, comprehended under the denomination of *libertinæ*, who were not expected to obey the rules by which maids and matrons of patrician birth were bound.

Girdles.—The girdle is of very great antiquity. That of the Greeks we find described in Homer. The Jews were enjoined to wear a girdle during the ceremonies of Easter. The Romans always wore a girdle to tuck up the tunic when they had occasion for exertion; this custom was so general, that those who went without a girdle, and allowed their gowns to fall loosely about them, were reputed idle, dissolute persons. For a soldier to be forbid to wear his girdle, was equivalent to degradation. The girdle was used by the ancients as a purse, or rather pocket. The belt of the Roman ladies, during the time of the emperors, answered the purpose of a corset. It was formed in front like a stomacher, and richly studded with jewels.

In the middle-ages, it was the custom for bankrupts and other insolvent debtors to put off and surrender their girdles in open court. The reason was that our ancestors used to place in the girdle all the articles, such as keys, purse, &c., that are now carried in the pockets, whence it became the symbol of the estate, goods, and chattels. This cession, or renunciation, as it was called, was no trifling matter, as it involved, for a noble, degradation from knighthood. The widow of Philip the Bold, Duke of Burgundy, in order to preserve the estates to her children, went through the cruel and disgraceful ceremony, by placing her girdle and keys on the coffin of the deceased, and declaring him bankrupt.

To take off the belt was a great point in the ceremony of rendering homage. It was only done in cases where *liege*-homage was rendered. The mooting of this point, which the Duke of Brittany refused to concede in doing homage to Charles VII., had well nigh produced a feud between the monarch and his powerful vassal.

The knight pronounced to be a felon and traitor, was despoiled publicly of his belt.

There was an ancient duty or tax laid in Paris, every three years, called the queen's girdle, which was intended for the maintenance of the queen's household. Vigènere supposes it to have been so called, because the girdle served for a purse; but he adds, that a like tax had been raised in Persia, and under the same name, two thousand years before, as appears from Plato and Cicero.

The Christians' girdle was instituted by Motavakel, caliph in the year of Hegira 235, to be worn by Christians throughout the East as a badge of their profession.

The Order of the Cordelière was instituted by Anne of

Bretagne, after the death of her first husband, Charles VIII., for widow ladies of noble families; it was placed round the escutcheons of their arms, and was also worn round the waist, with the ends hanging down. This order fell into disuse soon after the death of the founder.

The girdle of precious stones presented by Maria de Padilla to her royal lover, Don Pedro of Castile, was said to have been endowed with magical properties; and to this charmed belt was attributed the infatuation of the king for his mistress.

A girdle that will bear to future ages a better reputation, is the diamond one worn by the present Queen of Spain, Doña Isabel II., on the day of the murderous attempt made on her life by the curate Merino. The point of the assassin's dagger, striking on the diamond belt, slipped aside, and the blow, intended to be fatal, proved but a flesh wound.

Crowns.—Crowns were looked upon in the first ages as the insignia of the Divinity, and rather as a sacerdotal than a regal ornament. When the dignities and functions of priest and ruler were united in one person, the crown became the attribute of sovereignty. Thus we see the crown of the princes of Israel, the *pschent* of the Pharaohs, and the diadem of the sovereigns of the Anahuac, in nearly the same shape,—the shape of the episcopal mitre of our own day. The first mention in Holy Writ of the royal crown or diadem is in the book of Samuel, when the Amalekites brought Saul's crown to David.

Homer mentions kings with their sceptres, but not with crowns; as an insignia of royalty, the sceptre is more ancient than the crown.

The first diadems, worn only as the insignia of temporal power, were probably merely narrow bands or fillets, bound round the temples, and fastened at the back of the head, as we see them on the *dramatis personæ* of classic tragedy.

Crowns were subsequently worn as the insignia of victory, as indications of some special quality, as tokens of joy, pleasure, sorrow, &c. They were formed of branches of trees and of flowers; and as there was hardly a plant that had not some peculiar property or virtue, real or imaginary, so also was there hardly a divinity that had not a crown of some special plant attributed to it. The Master of the Gods, represented grasping his dread thunderbolt, was, strange to say, crowned with flowers; Juno had a wreath of vine leaves; the brow of the jolly god was more appropriately encircled with grapes, and ivy branches loaded with fruits and flowers. The Twin Brothers and the River Gods were crowned with reeds; Apollo with laurel; Saturn with ivy; Hercules with branches of poplar; the Graces and Minerva with olive-branches; the Hours with the fruits of each season; Ceres with wheat; &c.

Golden crowns were also offered to the gods. History mentions those sent to the Capitol by Attalus, and by Philip, King of Syria. Priests wore crowns of gold or of olive-branches, during the sacrificial ceremonies. The crowns of the flamens were of gold. Even the victims were crowned with pine or cypress. At funerals, crowns of laurel, olive, and sometimes lilies, were placed on the tombs. The Athenians took this custom from the Lacedemonians, and from Athens it passed to Rome. The moderns have had the good taste to continue this beautiful homage to the dead; and, in every European country, coronals of flowers are deposited on the tomb.

In Rome, on public days, the magistrates wore olive or myrtle crowns. From wearing a garland on festive occasions, it finally became the fashion at banquets to wear three; one on the crown of the head, another round the temples, and a third round the neck. The envoi by a Roman belle, to an admirer, of the partially faded coronal she had worn on the previous evening, was, according to the code of gallantry, a signal token of favor. The Latin poets have frequent allusions to this graceful *gage d'amour;* among others, Martial, in his distich to Polla:—

> "Intactas quare mittis, Polla, coronas?
> A te vexatas malo tenere rosas*."

Every flower composing these coronals had its signification.

It was only at sacrifices or banquets, however, that either sex could wear coronals of flowers.

But vanity was not long content with flowers; gold was gradually mingled with the roses, and the latter were finally supplanted by more costly, but more durable ornaments. Pliny tells us that P. Claudius Pulcher, consul in the year of Rome 569, introduced the custom of gilding the circlet of the crown, and covering with leaves of gold the branch of elm or the reed to which the flowers were attached. Ribbons fastened to the crown, and floating over the shoulders, were another innovation. These *lemnisci* were usually woven of silk and gold; sometimes they were embroidered with raised figures, and were very costly. One of Nero's courtiers, at a banquet which he gave to the Emperor, spent four million sesterces in silk crowns alone.

Bridegrooms wore a crown at their nuptials. The bride wore one of natural flowers when taken to her husband's house; and another of artificial flowers, of gold and precious stones, was placed on her head when she entered it.

Laurel, and finally gold, crowns were given as rewards of military valour. The Greeks gave crowns to those who had rendered the state some service. The first who bestowed this distinction on a Roman was the dictator A. Postumius, in the year of Rome 333. Having forced the

* "Wherefore, Polla, sendest thou me perfect crowns? I prefer the roses thou hast thyself faded."

camp of the Latins, near Lake Regillus, he granted from the spoils a gold crown to him who had contributed most effectually to their acquisition. The consul L. Lentulus granted one of five pounds weight to Servius Cornelius Merenda, after the capture of a Samnite city, in the year of Rome 472. The tribune Lucius Calpurnius Piso gave one of two pounds weight to his son, as a reward of the valour he had shown during a campaign in Sicily. But the father, whose economy had gained him the surname of *Frugi*, neither impoverished himself nor the state by his munificence to the youth; he allowed him the value of the crown, to be taken from the paternal heritage, by a clause in his will.

Military crowns were of several kinds. There were civic, mural, obsidional, and naval crowns. The obsidional was given by the besieged to their governor or commander, when he had forced the enemy to raise the siege. It was made of grass gathered within the walls of the town. The mural crown was the reward of him who, in an assault, had been the first to leap the walls of the besieged city, or to enter at the breach: it was of gold, and surmounted by towers. The civic crown was also of gold, and rewarded him who had saved the life of a citizen in battle. The naval crown was of gold, and adorned with naval emblems of the same metal; it was awarded to him who first boarded an enemy's vessel during an engagement. The castrensian crown was given to him who first forced the entrenchments of the foe: it was of gold, and presented the figure of a forced palisade.

In the year of Rome 467, Sicinius Dentatus, one of the bravest soldiers in the Roman army, alleged, in support of some demand he was making, that he had won fourteen civic crowns, one obsidional, and three mural; also eighty-three gold collars, and sixty bracelets of the same metal. Many such soldiers would have beggared the treasury.

The triumphal crown was the reward of the general

who obtained one or more signal victories. At first, it was of laurels, but subsequently of gold. In process of time, in lieu of one worn by the victor, a number were borne before him. According to Livy, two hundred and thirty-four gold crowns were paraded before Scipio Africanus; and Appian counted two thousand two hundred and eighty-two in the triumph of Cæsar. On these crowns were represented the chief exploits of the conqueror.

Claudius, in his triumph over Britain, had, among his crowns of gold, one of seven hundred pounds weight, furnished by Hispania Citerior, and one of nine hundred pounds, furnished by Gallia Comata.

The Rhodians sent every year a gold crown to Rome in token of friendship.

The medals of the Roman emperors show us four different crowns: the laurel crown, which Cæsar was so pleased that the senate had conferred upon him the right of wearing, because it concealed his baldness; the radiated crown; the crown adorned with gems or pearls; and a sort of bonnet, such as the princes of the empire sometimes have on their coat of arms.

The Roman emperors of Cæsar's family wore no other diadem than the laurel wreath, in imitation of the founder. Heliogabalus was the first who bound a fillet of pearls round his temples; and this sort of crown was subsequently much worn, especially from the time of Constantine. Some present a double row of pearls; on others, the pearls are mingled with precious stones set in gold. When the northern nations had destroyed the Eternal City, and when the East alone remained to the imperial dignity, considered the chief among the temporal sovereignties of Europe, the emperors wore the crown in use at Constantinople, and called by the librarian Anastatias, the *spanoclista*, that is, closed at the top. This crown, which was surmounted by a gold circlet, is the one of which a Latin author said:—"The imperial crown is the circle of the earth; it indicates universal power."

The pope wears a tiara, or triple crown, as indicative of his ecclesiastical and temporal power. Pope Hermiadas assumed the first, Boniface added the second, and Innocent XXII. the third.

In the middle-ages, the emperors claimed three crowns: one of silver at Aix-la-Chapelle, as kings of Germany; one, *soi-disant*, of iron at Milan, as kings of Lombardy; and one of gold at Rome, as emperors.

The Hungarian crown, worn at their accession by the emperors of Austria, as kings of Hungary, is the identical one worn by Stephen, eight hundred years ago; and from 1799 to the advent of his present majesty, it has been guarded day and night by two keepers. It is of pure gold, and weighs nine marks six ounces (fourteen pounds). The gems that adorn it are fifty-three sapphires, fifty rubies, one emerald, and three hundred and thirty-eight pearls. No sovereign of Hungary is legally invested with power and dignity until this diadem has been on his brow. It is shown to the populace three days before and three after the coronation. There do not appear to be any diamonds in the crown of Hungary.

Towards the tenth century, kings, dukes, marquises, and earls, assumed a crown or golden circlet, betokening absolute power, with the condition of simple homage to the suzerain.

The kings of the first race in France wore crowns of four different kinds:—a bandeau of pearls, with narrow bands hanging over the shoulders; another similar to the spanoclista of the Emperors; the third like the mortar-shaped cap of the president of the ancient French parliament; the fourth in the shape of a sugar-loaf hat, with a large pearl on the apex. The kings of the second race are usually represented with a double row of pearls, or an imperial mitre surmounted by a cross. The third race are seen with only one kind of crown, a golden circlet enriched with precious stones, and surrounded by fleurs-de-lis. The imperial crown, which has been worn since 1498, was adopted by Francis I., that he might not seem inferior in the insignia of sovereignty to his rival Charles V.

The famous iron crown of Lombardy is in reality of pure gold. It derives its name from a narrow iron circlet, said to be made of one of the nails with which the Saviour was crucified, which is enclosed in the interior of the crown. When Theodelinda, widow of Antharis, king of the Lombards, bestowed her hand and kingdom on Agilulph, duke of Turin, she presented him with this precious crown, since worn by all who became masters of Italy. It was kept in the treasury of the monastery of Monza, near Milan.

The first crown worn by a king of the Franks was the one sent to Clovis by the Emperor Anastatius, together with the diploma of consul. The crown was of gold adorned with gems. The Frankish prince, repairing to the basilica of St. Martin, was there ceremoniously invested with the consulary chlamyd and tunic, and crowned with the diadem. Thus arrayed, he mounted his horse, and, followed by a brilliant cortége, he proceeded from St. Martin's church beyond the walls, to the cathedral of Tours, scattering on his way handfuls of gold and silver to the delighted crowd. The mode of investiture of the royal dignity, or rather, the proclamation of a new chief, had hitherto consisted in the soldiers raising the elected warrior on their shields.

The Peruvian Incas wore also an insignia of sovereignty around their temples; it was not, however, of gold, but consisted, like the pschenck of the Egyptian Pharaohs, of a fillet of tasselled fringe of a yellow color. The Peruvian ornament was made of the fine threads of the Vicuna wool.

The crown of the ancient Mexican sovereigns was in the shape of a mitre of gold, and elaborately adorned with precious stones and feathers.

The ancient Persian monarchs wore as the insignia of sovereignty a purple and white tiara called *cydaris*. There is no crown in the Persian regalia of the present day, unless the cap enriched with jewels worn by the king be so called. Tavernier, who witnessed the accession of several oriental princes, thus describes the investiture :—

"I was present at the accession of Cha Sephi I. and of Cha Abbas II., but saw no crown placed on the monarch's head on either occasion; nor is the ceremony of a coronation used in any part of Asia. The principal mark of investiture, both in Persia and Constantinople, is the girding on of the sabre. A cap, covered with the richest jewels in the treasury, is placed upon the new sovereign's head; but this cap has no resemblance to a crown. The same ceremony of the sabre and cap is used with the Great Mogul, the king of Visapour, and the king of Golconda."

Among the toys carried as indispensable necessaries, snuff-boxes were once greatly in vogue; gold, silver, enamel, precious stones, and exquisite paintings, gave value to these trinkets. The snuff-box was, in the last century, carried by women as well as men, and that and the *bonbonnière*, a sugar-plum box, were usually very expensive appendages to the beaux and belles before the first French Revolution. No one, however, seems to have indulged this fancy to such an extent as king Frederick II. of Prussia, who had as many as fifteen hundred snuff-boxes, many of which were extremely rich.

Shoe-buckles, in the reign of the sixteenth Louis, were so large that they covered all the instep.

CHAPTER IV.

EAR-RINGS.

Their Antiquity.—Juno's Ear-rings.—Penelope's Ear-rings.—Ear-rings of the Egyptian Ladies.—Eve's Ears bored.—Fatal Ear-rings of the Israelites.—Arab Saying.—Weight of Jewish Ear-rings.—Ear-rings among the Greeks and Romans.—Extravagance of Roman Belles.—Different kinds of Ear-rings.—Ear Doctresses.—Title of the Emperor of Astracan.—The Ear-ring an Insignia of Knighthood.—Ear-jewels of the Chola Girls.

"Far beaming pendants tremble in her ear,
Each gem illumined with a triple star."
POPE'S *Homer's Iliad.*

THOUGH among civilized nations the nose jewel is discarded,—and with good reason, as the ornament must be very inconvenient,—the ear-ring has maintained its ground triumphantly to the present day. With the ladies of the present day it is considered an indispensable adjunct to the toilet.

Ear-rings are ornaments of the most remote antiquity. In Homer, we find Juno placing pendants in the lobe of her ears. Ear-rings, too, are among the rich gifts presented by Eurydamas to Penelope.

Among the Athenians, it was a mark of nobility to have the ears bored; and among the Hebrews and Phœnicians, it was (for men) an indication of servitude.

In the ancient tombs of Egyptian kings, agates, chalcedonyxes, onyxes, cornelians, shaped like perfectly round pearls and beautifully polished, have been found; they had evidently been worn as ear-jewels.

The ear-rings worn by the Egyptian ladies were round, single hoops of gold, from one inch and a half, to two inches and one-third, in diameter, and frequently of still greater size, or made of six single rings soldered together. The ancient ear-rings seen in the sculptures of Egypt and Persepolis, are generally of a circular form. Such, probably, was the round "agil" of the Hebrews. Among persons of high or royal rank, the ear-ring was sometimes in the shape of an asp, the body of which was of gold, set with precious stones.

Silver ear-rings have also been found at Thebes,—either plain hoops, like modern gold ones, or simple studs. The modern oriental ear-rings are more usually jewelled drops or pendants than circlets of gold.

The Rabbis assert that Eve's ears were bored when she was exiled from Eden, as a sign of slavery and submission to man, her master. If so, the slaves have since found a way to make their masters atone for this humiliation; the latter must pay dearly the diamond badges of their wives' servitude. Since then, not money alone have these pretty baubles cost; blood has been poured forth in torrents to procure them for some capricious fair one, while the sacrifice of them has, at other times, been attended with the most fatal results. The golden calf was made entirely from the golden ear-rings of the people,—probably the same they had borrowed of the Egyptians, and neglected to return,—and three thousand men paid with their lives the unworthy use to which the jewels were put.

We find also, that the ephod, made of the ear-rings of the princes of Midian, "became a snare unto Gideon and to his house."

Among the Arabs, the expression, *to have a ring in one's ear*, is synonymous with *to be a slave*. When one man submits to the will of another, he is said to have placed in his ear the ring of obedience.

Ear-rings play an important part in the Old Testament. The name in Hebrew denotes roundness, and therefore, was applicable to any kind of ring. Another word, *nezem*, appears to have denoted sometimes a nose jewel, and sometimes an ear-ring.

If we have no positive information as to the shape of the ear-jewels worn by the Israelite ladies, we can form some idea of their weight from that single one given to Rebekah, which was of gold, and weighed half a shekel, (about a quarter of an ounce).

Ear-rings of certain kinds were anciently, and are still in the East, regarded as talismans and amulets. Such, probably, were the ear-rings of Jacob's family, which he buried with the strange gods at Bethel (Gen. xxxv. 4). St. Augustine speaks strongly against ear-rings that were worn as amulets in his time (Epist. 75, *ad Pos.*). Schrœder, however, deduces from the Arabic, that the amulets were in the form of serpents, and similar, probably, to those golden amulets of the same form which the women of the pagan Arabs wore, suspended between their breasts, the use of which was prohibited by Mahommed.

The use of the ear-ring appears to have been confined to the women, among the Hebrews. That they were not worn by men, is implied in Judges: "And Gideon said unto them, I would desire a request of you, that ye would give me every man the ear-rings of his prey. (For they had golden ear-rings, because they were Ishmaelites.)" viii. 24. Yet when gold is needed to make the calf, Aaron bids the people, "Break off the golden ear-rings which are in the ears of your wives, of your *sons*, and of your daughters." (Exod. xxxii. 2.)

The men of Egypt also abstained from the use of ear-rings, though the sculptured heads on the ancient monuments prove that they were extensively worn by men in other nations.

The Grecian ladies wore ear-rings as well as finger-rings, adorned with gems; and the Roman belles, who derived their fashions from the East, were not backwards in adopting this one.

A guest at the feast of Trimalcyon complains, that the jewels of his wife have wasted his patrimony. "Should I have a daughter," he exclaims, "I will cut off her ears at her birth to spare myself first, and my son-in-law afterwards, the ruinous expense of ear-rings."

It was more especially in their ear-rings that the vanity and ostentation of the Roman ladies was displayed, probably because, of all their ornaments, these were the most in view. Pearls were the most expensive ear-rings. The fashion at first was of one pear-shaped pearl in each ear; these were called *uniones*, and usually cost enormous sums. It is probable that they were not always worn of like size and shape in both ears, at least it would seem so, from the fact that Julius Cæsar presented to the mother of Brutus a single pearl, and not a pair,—and that single pearl cost six million sesterces!

The *uniones* were succeeded by the *crotalii*, formed of two, three, and sometimes four, large pearls in a row, that rattled at every motion of the head; hence their name. The price of these ornamental *rattles* was so exorbitant, that Seneca indignantly exclaims:—"One pearl in each ear no longer suffices to adorn a woman; they must have three, the weight of which ought to be insupportable to them. Women, in their madness, deem that their husbands would not be sufficiently tormented, if they did nor wear in each ear the value of three fortunes."

The ear-rings of the Empress Poppæa were worth three million francs; those of Cæsar's wife, six million. As only matrons of the highest rank and greatest wealth could wear such ornaments, they were considered equivalent to the lictors that preceded the vestals, as indicative of the respect due to the wearer; and the name of the ear-ring expressed this.

The forms of ear-rings were numerous. There was the *bulla* bubble, perhaps so called on account of their shape and lightness, being made of very thin gold.

The *caloiea*, large ear-rings set with a green precious stone; perhaps an emerald.

The *caryotæ*, in the shape of small green nuts, as the name indicates.

The *centauri*, adorned with little golden figures of centaurs.

The *hippocampus*, to which hung figures of horses, or of a fish very common in the Mediterranean, and known as the marine horse.

The *rotulæ* were in the shape of small wheels.

The *stalagmii*, in the shape of golden pearls.

What the *triglenæ*, so famous as one of the ornaments of Juno, and as being presented to Penelope, were, no *savant* has as yet explained. Some ear jewels were in the shape of little tripods, and bore that name.

There were women in Rome, whose sole occupation was the healing of the ears of the belles who had torn or otherwise injured the lobes with the weight of their pendants; these special doctresses were called *auriculæ ornatirci.*

Alexander Severus forbade men wearing ear-rings.

In Greece, children wore an ear-ring in the right ear only.

The importance given to the ear-ring among East Indian princes, is evidenced in the title of the Emperor of Astracan, who styles himself, "Emperor of Astracan, Possessor of the White Elephant, and the Two Ear-rings, and in virtue of this possession, legitimate heir of Pegu and Birma, Lord of the Twelve Provinces of Bengal, and the Twelve Kings who place their heads under his feet."

The ear-rings worn by the East Indians, both men and women, are often of enormous size. It is the fashion to lengthen out the ear, and to enlarge the hole, by putting in rings the size of saucers, set with stones.

In South America, the ear-ring played also an important part. Among the Incas, it was a badge of knighthood. The sovereign himself condescended to pierce, with a golden bodkin, the ear of the aspirant deemed worthy of that honor; the bodkin was suffered to remain there until the aperture was sufficiently large to admit of the enormous ring which distinguished this order of nobility. The ornament worn by the monarch himself was so heavy, that the cartilage was distended nearly to his shoulder.

The fashion of enormous ear jewels seems to be hereditary in Peru, at least with a certain class; those of the Chola girls (mixed descendants of Peruvians and Spaniards) are so ponderous as to require being supported by a golden chain, which passes over the head.

Keys were formerly used in England as an ornament to the ear; this fashion is referred to in the fifth act of *Much Ado About Nothing*, wherein Dogberry exclaims: "They say he (Conrade) wears a key in his ear, and a lock hanging to it." In Shakspeare's time, there lived in London a smith, called Mark Scaliot, who, in 1578. manufactured and exhibited as specimens of his skill, a lock of iron, steel, and brass, of eleven several pieces, and a pipe-key, all weighing but one grain of gold; he also made a chain of gold of forty-three links, which chain, being fastened to the lock and key, and put about a flea's neck, the flea drew them with ease. Probably, this Mark Scaliot was the manufacturer of the lock-and-key trinkets that fashion placed in the ears of the beaux and belles of that day.

The pictures of Henry II. and Henry III. of France, and their courtiers, show that ear-rings were then worn by men. The fashion prevailed among the beaux of the court of Queen Elizabeth, and Shakspeare's ears were similarly adorned. The gallants of the Directoire wore hoops in their ears; and they are still frequently seen among the lower classes, especially sailors, on the continent.

CHAPTER V.

BRACELETS.—ARMLETS.

The Armlet a Token of Sovereignty in the East.—Worn by Men.—Antiquity of the Bracelet.—Egyptian Bracelets.—Not an Ancient Fashion with the Greeks.—Mentioned in Holy Writ.—The Bracelet among the Moderns in the East.—The Armillæ of the Romans.—Armlets of the Sabines.—The Bracelet not worn by Girls.—Different kinds of Bracelets worn in Rome.—Armlets of the Gauls and Saxons.—Used to render Contracts binding.—Celebrated Armlets.

> "Tu quoque et auratos Eriphyla, lacertos,
> Dilabsis nusquam est Amphiaraus equis."

The word bracelet, from the Latin *brachiale*, is applicable, as its etymology indicates, to any circlet worn on the arm; but it generally designates only the ornaments placed on the wrists: that which adorns the arm above the elbow is termed an armlet.

There is also this difference, that, in the East, *bracelets* are generally worn by women, and *armlets* by men. The armlet, however, is in use among men only as one of the insignia of sovereign power. Such probably was the royal ornament which the Amalekite took from the arm of the dead Saul, and brought with the other regalia to David. There is little doubt that this was such a distinguishing band of jewelled metal as we still find worn as a mark of royalty from the Tigris to the Ganges. Still, the bracelet was not the attribute of a king only; Judah, who was the head of a tribe, wore them also. The kings of Persia presented bracelets to all the ambassadors from foreign courts.

The Egyptian kings are represented with armlets, which

were also worn by the Egyptian women. These however, are not jewelled, but plain or enamelled metal, as was in all likelihood the case among the Hebrews. The bracelet and armlet are of the highest antiquity. Some Egyptian bracelets are several centuries older than the most ancient Greek monuments. They were of different colors; of beautifully wrought gold, set with precious gems, or adorned with very bright and fine enamel.

Bracelets came into fashion among the Greeks much later than rings. The invention and use of this ornament could only originate with a people whose arms were bare; and the ancient Greeks, who had their costumes from Ionia and the East, and wore long sleeved tunics, probably never thought of adopting the bracelet, until they changed the style of their dress for the Dorian.

Bracelets are mentioned in several passages of the Scriptures; and we may judge of their weight and value from the fact, that those presented to Rebekah weighed ten shekels (five ounces). Sir John Chardin tells us, that ornaments as heavy, if not heavier, are worn now by the Eastern women. They are there placed one above the other, until the arms from wrist to elbow are covered with them. Many of these are so massive and heavy, as to be more like manacles than ornaments. The materials vary according to the condition of the wearer, but it seems to be the rule, that bracelets of the meanest descriptiou are better than none. Among the higher classes, they are of mother-of-pearl, of fine flexible gold, and of silver, the last being the most common. The poorer women use plated steel, horn, brass, copper, beads, and other materials of a cheap description. They are sometimes flat, but more generally semi-round, or circular, except at the point where they open to admit the hand, where they are flattened. They are frequently hollow, giving the show of bulk (which is much desired) without the inconvenience. Bracelets of gold twisted rope-wise, are those now most in use in Western Asia; though it cannot be determined to what extent this fashion may have existed in ancient times.

Bracelets among the Romans were at once a mark of honor and a token of slavery; in the latter case, they were merely iron or brass bands. The gold bracelet, the armilla, was at first given as the reward of military valor, conferred by princes and generals. Afterwards they became arbitrary, and were worn at pleasure. An ancient inscription, quoted by Gruter, represents two armlets with these words: *L. Antonius, L. F. Fabius, Quadratus Donatus Torquiabus, Armillis ab Tiberio Caesare bis.*

According to Titus Livy, the Sabine warriors wore very heavy bracelets on their left arms. These glittering ornaments tempted the unwary Tarpeia to betray the Roman citadel to the foes, on condition they would give her their gold armlets, or, as she expressed it, what they carried on their left arms. The conquerors paid the treason and punished the traitoress by hurling, not only the coveted bauble, but the massive buckler, also worn on the left arm, at the head of the vain wretch. If the bracelets of the Sabines were as heavy as some of those that are in the cabinets of antiques, they were quite sufficient of themselves to have crushed Tarpeia, without the addition of the bucklers. The ancient massive bracelets, in the shape of serpents wound several times round the arm, that are still extant, were probably masculine ornaments.

Bracelets were never worn among the Romans by unmarried females, at least not until they were betrothed. They made up afterwards for the privation they had endured when girls; according to Petronius Arbiter, the satirist, some of them wore gold bracelets weighing six and a half, and even ten pounds.

Among the lava-buried relics of Pompeii, was found a lady with two bracelets on one arm.

In the time of Pliny, men wore bracelets of gold, surnamed *Dardanian*, because brought from Dardania.

The Emperor Maximilian, who succeeded Alexander Severus, and who was eight feet and one inch in height, wore his wife's bracelet as a thumb ring.

The ancient Roman bracelet was of different shapes. Those of women were sometimes in the shape of a serpent, or in that of a rope, or round braid ending in two serpents' heads. Sometimes they were placed above the elbow, and sometimes on the wrist: the wrist-bracelet was called by the Greeks *perecarpia*. On the statue of Lucilla, the wife of the Emperor Lucius Verus, there is a bracelet that forms three circles round the wrist. In Capitolinus, the bracelet is twice called *dextrocherium*; in the great inscription of Isis, it is called *lucialium*.

The Gauls wore massive gold armlets and bracelets.

Plutarch, Xenophon, Herodian, Isidorus, all allude to bracelets as military rewards; the Draconarii, or standard-bearers, wore them. Sicinius Dentatus boasted of his sixty gold bracelets won by his valor; and Aulus Gellius, the Roman Achilles, had obtained more than a hundred and sixty.

When Clovis bribed the *leudes* of Ragnachaire, King of Cambrai, to betray their lord, the price of treason was to consist of bracelets and hauberks of gold. The Frankish prince was as false as those he corrupted: the gold turned out to be copper gilt.

The emblems of authority among British kings were gold bands worn on the neck, arms, and knees. Dion Cassius, and other writers, describing the dress of the warlike British queen, Boadicea, inform us, she wore a chain of gold about her neck, and bracelets on her arms.

In the *Saxon Chronicle*, under the year 965, the Saxon monarch Edgar is called the bestower of bracelets, the rewarder of heroes. With the Norwegians, Gauls, Celts, and Saxons, the bracelet was the reward of valor.

According to William of Malmesbury, the Saxons, just before the conquest, loaded their arms with massive bracelets of gold.

Armlets were worn by the Normans when they invaded Gaul. The security of the roads, and the clearance made of robbers by the strict laws of Rollo, the great chief of these invaders, is exemplified by an incident in his life. One day after hunting, as he was taking his repast near a brook in a forest in the vicinity of Rouen, he hung his golden armlets on the branches of an oak, and there forgot them. The jewels remained there three years, no one daring to touch them.

Bracelets, like crowns, chains, and other ornaments, were made votive ornaments. Matthew Paris, in his description of King Henry III.'s visits to St. Alban's Abbey, in 1244, (where he remained three days each time,) says, that on the last day he gave a rich pall or cloak at the high altar, and three bracelets of gold at the shrine of the saint.

Northern nations used to swear on their bracelets, to render contracts more inviolate.

Bracelets have been worn by both sexes, not only in various countries of the East, but also by different savage tribes of Oceania, who manufacture very pretty ones with the bark of some trees, with feathers, shells, beads, &c.

In modern times, the most celebrated armlets are those which form part of the regalia of the Persian kings, and which formerly belonged to the Mogul emperors of China. These ornaments are of dazzling splendour, and the jewels in them are of such large size and immense value that the pair are estimated at two hundred thousand pounds. The principal stone of the right armlet is famous in the East by the name of the Devia-e-nur, or Sea-of-light. It weighs one hundred and eighty-six carats, and is considered the diamond of finest lustre in the world. The principal jewel of the left armlet, although of somewhat inferior size (one hundred and forty-six carats) and value, is renowned as the Tah-e-mah, Crown-of-the-moon. The imperial armlets, generally set with jewels, may also be observed in most of the portraits of the Indian emperors. The famous Koh-i-noor was worn by Runjeet Singh, set in an armlet.

Bracelets constituted a part of the regalia of the ancient Mexican and Peruvian sovereigns.

CHAPTER VI.

RINGS.

Earliest Mention of Rings.—Mistaken Idea of the Ancients.—The Ring a Symbol of Omnipotence.—Rings among the Hindoos.—In Holy Writ.—Seal-Ring among the Hebrews.—No mention of Rings in Homer.—Seals, but no Seal-Rings, among the Ancient Americans.—How worn among different Nations.—Rings among the Romans.—The Iron Ring.—Fable of Prometheus.—The Gold Ring.—By whom worn.—Dissensions it occasioned.—Edicts with regard to it.—Absurd length to which the Fashion was carried.—Rings on every Joint.—Winter and Summer Rings.—The Pugilists' Ring.—Rings of Lawyers and Orators.—Hired as aids to Eloquence.—The Dactyliomancia; Charmed, Consecrated, and Hallowed Rings.

> "Et tamen hoc ipsis est utile: purpura vendit
> Caussidicum, vendunt amethystina; convenit illis
> Et strepitu, et facie majoris vivere census.
> Sed finem impensæ non servat prodiga Roma
> Fidimus eloquio! Ciceroni nemo ducentos
> Nunc dederit nummos, nisi fulserit annulus ingens."—JUVENAL.

> "Pessimum vitæ scelus fecit, qui id primus induit digitis. Nec hoc quis fecerit traditur."—PLINY.

THOUGH we do not agree with Pliny in his condemnation of this pretty ornament and its inventor, we are not one whit better informed as to who was the first wearer of it. Its existence, however, among the relics of the most remote ages in all nations, induces the conviction that, of all ornaments, this was the first worn. The earliest mention we have of rings shows them to have been used as symbols, tokens of trust, insignia of command, badges of rank and honor, pledges of faith and alliance, and also tokens of servitude.

In the religious system of Zoroaster, Ormuzd was the Good Principle, the adversary of Ahrimanes, the Evil Principle. His name, in the Zend language, is Ahovera

Mazda, which means, very wise sovereign. He is represented in ancient sculptures holding a ring in one hand as the symbol of omnipotence.

The Greeks called this little circlet by a name derived from the Greek word δάκτυλος, finger; the Romans, from the word *ungula*, nail, as it was primitively worn by them on the first joint of the finger, and near the nails.

It is said that rings were originally worn in the East to prevent the fingers from growing large. If so, the idea was as great a mistake as that of the Chinese with regard to the foot; the constant wearing of rings certainly reduces that portion of the finger thus covered, but at the expense of the joints, which grow larger in proportion as the intermediate space shrinks.

Several of the Egyptian rings in the Museum of the Louvre date from the reign of King Moeris. In the Egyptian gallery in the British Museum, there is a very interesting gold ring of the Ptolemaic or Roman period, with figures of the deities Serapis, Isis, and Horus. Here are also several signets set with amulets or scarabaei (the sacred beetle), and others bearing the prenomen of Thothmes III., and Rameses III. or IX., with iron ones of the Greek period. One of the oldest rings extant is that of Cheops, the builder of the Great Pyramid, which was found in a tomb in the vicinity of the time-baffling monument. It is of gold, with hieroglyphics.

The fashion of rings was transmitted by the eastern nations to the Greeks, and by these to the Romans. All the Hindoo-Mogul divinities of antiquity, as well as those of the present day, are loaded with rings. The statues of the gods in the island of Elephanta, supposed to be of the highest antiquity, that is, of a date ten centuries earlier than our own, though they have no garments on, are adorned with head-gear, necklaces, ear-rings, finger-rings, belts, and also their various attributes.

The antiquity of rings is shown by many passages of

the Scriptures. When Pharaoh committed the government of all Egypt to Joseph, he took the ring from his finger and gave it to him as an insignia of the command he vested in him. In like manner did Ahasuerus to his favorite Haman; and to Mordecai, when the latter, by a court intrigue, was made to succeed Haman in all his dignities.

The finger-ring was, primitively, more especially used among the Hebrews, as a stamp of authenticity to letters and other instruments in writing. In the first Book of Kings, Jezabel "wrote letters in Ahab's name, and sealed them with his seal." Repeated instances of the use of the seal are found in Holy Writ. In the Book of Jeremiah there is a very remarkable passage, in which all the formalities attending a Jewish purchase, including that of the seal, are given, viz.: "And I bought the field of Hanameel, and weighed him the money, even seventeen shekels of silver. And I subscribed the evidence, and sealed it, and took witnesses, and weighed him the money in the balances. So I took the evidence of the purchase, both that which was sealed according to the law and custom, and that which was open."—c. xxxii.

In the East, the seal took place of the signature, or was added to it to give it validity, as may yet be seen in ancient Babylonian documents. The writings were made in the presence of the sovereign, his seal was affixed to them, and they were forthwith despatched. The impression of the monarch's signet-ring gave the force of a royal decree to any instrument. "*Write ye also for the Jews, as it liketh you, in the king's name, and seal it with the king's ring: for the writing which is written in the king's name, and sealed with the king's ring, may no man reverse.*" Hence the delivery or transfer of it to any one gave the power of using the royal name, and created the highest order in the state. Rings, being so much employed as seals, were called *tabaoth*, which is derived from a root signifying to imprint, and also to seal. They were commonly worn as ornaments on the fingers, usually on the little finger of the right hand.

The Greeks, however, according to Pliny, made no use of rings in the time of the Trojan war. His reason for thinking so is, that we find no mention of them in Homer, either as seals or otherwise. When letters were to be sent, they are described as being tied up and the strings knotted.

Disseminated throughout the immense territory of that portion of the new continent once possessed by Spain, are still found quantities of seals on hard stones, grünstein, lava, obsidian, agates, serpentine, silex, and terra-sigillata, which the inhabitants fastened to the wrist or arm, and which they used in their contracts, in lieu of signatures. The seals of kings or high dignitaries were of precious or fine stones, or of gold; and bore, in relievo, figures of animals, arabesques, emblems, &c. But neither Mexicans nor Peruvians wore their seals set in rings, though these were often adorned with engraved stones.

At a later date, the use of rings is found among the ancient Chaldeans, Babylonians, Persians, Greeks, and, subsequently, the Romans. According to Herodotus, every man in Babylon wore a seal-ring, and the fashion probably extended from thence to the Medes and Persians.

The most precious stones were used for seals. The ancients forbade the engraving of the figure of God on seals. But in process of time this was little regarded, and the Egyptian and other deities, the effigies of sovereigns, of heroes, friends, and ancestors, figures of animals, the constellations, parts of the human body, etc., were also engraved.

We find ancient Mexican rings and seals set with fine or precious stones, on which the constellation of Pisces is cut. The Mexicans, as well as Hebrews, awaited their Messiah, or the Crusher of the Serpent, during the conjunction of Jupiter and Saturn, in the same zodiacal sign of the Fishes,—the protecting sign of Syria and Palestine.

The manner of wearing rings has been very much diversified according to the age and the nation. The Hebrews placed the ring on the right hand; the Greeks wore the ring on the fourth finger of the left hand; the Gauls and Britons, on the third finger of that hand. Among the Romans, according to Pliny, the left hand, as being "the least noble," brought the ring into fashion. The Hindoos,—in fact, all the Asiatics, the Peruvians, and Mexicans, carried the love of rings to a greater length, for they wore them not only on all their fingers but on their toes; many savage tribes in Asia, Africa, and America, wore them also in their nostrils, cheeks, and chin.

All kinds of ornaments having been interdicted by some of the Grecian legislators, in order to make the prohibition more effective, an exception was made in favor of women of ill-fame, these being permitted to wear gold and silver trinkets. This measure was supposed to rob the precious metals of their value. Honored matrons wore but an iron ring. That the iron ring was a souvenir of slavery, we find in the fable of Prometheus, who, when freed by Hercules, was still compelled to wear an iron ring, in remembrance of the links of the chain that had bound him to Mount Caucasus. A ring made of this base black metal, might have been looked upon by sensitive and refined beings—which, for their own happiness, it is to be hoped the Spartan dames were not—as a stern reminder of an ungilded servitude.

The elegant and luxurious Athenians imposed no such restrictions on the fair sex, and we have descriptions of the toilet-tables of contemporaries of Aspasia, that prove the variety and richness of the jewels then worn.

It is not known who first introduced the wearing of rings among the Romans. It was Tarquin the Elder who first made them insignia of knighthood, and they also became the rewards of valor. But even then, no costlier metal than iron was used. The statue of Romulus in the Capitol had no ring, neither had the other statues, not even excepting that of Lucius Brutus; yet those of Numa Pompilius and Servius Tullius, had. That rings should not have been placed on the statues of the Tarquins, was the more surprising, as these princes were from Greece, whence the fashion had doubtless come to Rome.

It was long before even senators wore gold rings. The state bestowed them only on such as were sent on embassies to foreign nations; probably, with a view to add to their dignity among people living in semi-barbaric pomp. But, unless given by the state, the gold ring was not worn in the earlier days of the Republic. The victor honored with a triumph, wore no other ring than an iron one, similar to that of the slave who held over his head the crown of Etruscan gold. Thus did Caius Marius make his entrance into Rome, after his conquest over Jugurtha. It was not until after his third consulship that he ventured to take the gold ring. Those who had obtained this precious circlet, only wore it on public occasions; at home they resumed their iron one.

The *annulus sponsalium* is of Hebrew origin. The custom was adopted by the ancient Romans. Before the celebration of the nuptials, the betrothals were celebrated, very much as they are at the present day on many parts of the continent. At the conclusion of the feast, the bridegroom placed, as a pledge, on the fourth finger of the bride, a ring. The fourth finger was preferred, from a belief that a nerve reached thence to the heart. A day was then named for the marriage.

The ring presented to the betrothed maiden was still, in the age of Pliny, an iron one; a loadstone was set in place of a gem. It indicated the mutual sacrifice made by the husband and wife of their liberty; the magnet indicated the force of attraction which had drawn the maiden out of her own family into another.

Monsieur Latour St. Ybars alludes to this custom in his tragedy of *Virginia;* and the passage in which the indignant maiden rejects with scorn the richer gifts of a seducer, and speaks of the iron ring given to her by her affianced husband, is so beautiful, we may be excused for quoting it here.

> "Alors qu' Icilius ne m'a jamais offert
> Pour gage de sa foi que cet anneau de fer,
> Claudius, sans respect pour l'amour qui m'anime,
> Par cet appât grossier croit m'entraîner au crime;
> Et ces ornements vils qu'il m'ose présenter,
> Sont faits de ce métal qui sert pour acheter!
> Va rendre à Claudius tous ces dons, et sur l'heure
> Les présents de cet homme ont souillé ma demeure;
> Et ce serait blesser notre honneur et nos dieux
> Que d'y porter la main, que d'y jeter les yeux."

A little more than a century later, we find the base metal discarded. Tertullian, and Isidore, Bishop of Seville, speak of the *annulus nuptialis, sponsalitius,* as being of gold. Men then, as yet, wore but two rings.

It does not appear that the use of rings became common before the time of Cneius Flavius, son of Annius. Having acquired the confidence of the people, he was appointed a *curale edile* with Q. Anicius, who, a few years before, had been an enemy to Rome, to the exclusion of C. Postilius and of Domitius, whose fathers had been consuls. He was at the same time made a tribune of the people. This excited such indignation, he being a freedman, that the patricians deposited their rings, and even their *phaleres* (collars). The functions of a military tribune, which lasted six months, conferred the right of wearing the gold ring, a privilege of which the patricians were exceedingly jealous.

Rings must have been in general use among the higher classes, after the second Punic war, since Hannibal sent three bushels of knights' rings to Carthage.

The flamen of Jupiter could only wear a hollow ring of very thin gold.

The rings presented to friends and relations, on birthdays, bore symbolic signs or mottos.

The golden ring on the fourth finger, designated a knight, and distinguished the second order, as the laticlave designated the senator.

In the year of Rome 775, during the consulate of C. Asinius Pollion and of C. Antistius Vetus, a decree fixed the right of wearing the ring. An apparently trifling incident gave rise to it. C. Sulpicius Galba, seeking, when yet young, to acquire influence with the sovereign, by the exercise of a rigorous surveillance over tavern-keepers, complained to the senate that delinquents usually escaped punishment by virtue of the privilege attached to their ring. Whereupon, it was decreed, that no one should have the privilege of wearing a ring, unless he, his father, and his grandfather, all freeborn men, had possessed four hundred thousand sesterces, (eighty-four thousand francs, or three thousand three hundred and sixty pounds sterling,) in landed property, and had, in accordance with the *lex Julia* on theatres, the right to sit on the fourteen rows of seats.

The assumption of this honorable distinction became so frequent, that Proculus, one of the knights, denounced to the Emperor Claudius, then Censor, four hundred persons, charged with wearing it illegally.

The reign of Claudius witnessed another distinction, which consisted in a gold ring, on which was engraved the effigy of the sovereign. The right to wear this ring was obtained from the Emperor's freedmen. This privilege also gave rise to innumerable accusations of usurpation, which the advent of Vespasian put an end to; that emperor declaring that all subjects had a right to the image of the prince.

Rings at last became common ornaments, the right of all who could afford to wear them, and only more or less costly, according to the means of the wearer. So many, who could not afford a richer metal, wore gilt rings. Thus, that which had once been the insignia of the sovereign, the senator, the knight, the badge of good service done on the field, became common to the buffoon, the juggler, and the courtesan.

When Nero sang in the circus, he was attended by a band of beautiful children, whose flowing locks were perfumed, and who each wore a gold ring. Galba chose a number of young men of the order of the equites, to mount guard at his palace, in lieu of the soldiers who formerly performed this duty. It was particularly specified, that the new cohort should keep their gold rings, and bear a title designating their origin.

The lower classes wore iron rings, but these were set with low priced stones, such as unengraved agates and cornelians, or with colored glass and paste stones imitating gems. Manumitted slaves were presented by their masters with a robe, a cap, and a ring.

As a token of bargain, the ring was drawn from the finger.

In getting up pleasure parties, where each guest contributed his quota to the expense—what moderns call pic-nics—the Romans made a temporary exchange of their rings, as vouchers that they would fulfil their engagements.

The fashion of rings was finally carried to a most absurd extreme by both sexes. A ring was at first only worn on the forefinger; a second was afterwards placed on the fourth finger; then again, another on the third; until at last the thumb and all the fingers, except the middle one, were loaded. The most valuable was placed on the little finger, and was never used as a seal ring. Smaller rings were fitted to the second joint of the fingers.

Three rings were sometimes placed on the little finger.

Lucian speaks of a rich Roman, who wore sixteen rings —two on each thumb and finger, except the middle one,— which was held in a species of reprobation.

Seneca, speaking of the luxury of the Romans, says:— "Our fingers are loaded with rings; each joint is adorned with a precious stone."

"Luxury," says Pliny, "which has corrupted all things, and the influence of which is felt in this respect, in a thousand different ways, has introduced the fashion of adding to these rings precious stones of the most lively and brilliant *éclat;* the wealth of a whole family is worn on the finger."

There were winter and summer rings, designated by Juvenal as *aurum semestre, aurum æstivum, annuli semestres.* The weight and color of the rings was adapted to the season. The sardonyx, the cornelian, the rock-crystal, the hyacinth, were summer rings, as being deemed cooler and lighter gems; the heavier rings were worn in winter. Some of these, preserved in the cabinets of antiquaries, weigh as much as an ounce.

The Greeks had their weekly rings. As these rings were all cameos and intaglios, and were eagerly sought by men and women, it is not surprising that so many should have come down to us. Every rich Roman was a walking dactyliotheca.

Heliogabalus never wore the same ring twice.

Among the offensive weapons sometimes made use of by the ancient pugilists in their encounters, the enormous ring, weighing sometimes two ounces, inflicted a mortal blow.

Lawyers invariably wore a ring when pleading a cause, and when they owned none, they hired one for the occasion. This custom Juvenal repeatedly stigmatized:

> ...... "Ideo conducta Paulus agebat
> Sardonycho et que ideo plurisquam Cossus agebat
> Quam Basilus. Rara in tenui facundia panno."

Authors, who read their verses in public, did not neglect this aid to their eloquence, and Perseus has not failed to satirize these foppish poets.

The passion for rings was such, that the wealthy formed collections of these ornaments, to which the foreign name of *dactyliothecæ* was given, from two Greek words, signifying *case* or *box*, and *ring*. The name was subsequently given to collections of engraved gems, such as are now more properly termed *glyptothecæ*.

Another fashion, adopted by men, was that of rings bearing the head of Harpocrates, or of some Egyptian divinities—Isis, Horus, Serapis, &c.

The Romans had also their amulets, or magic rings, on which were engraved one or more stars, the head of Anubis, a sign of the zodiac, or a human foot.

Ancient Roman finger-rings may be seen among the Greek and Roman antiquities in the British Museum, in the Bibliothèque Impériale of Paris, and in many other cabinets of ancient gems in the European capitals.

But the ring mania was not content with considering the ring as an ornament, or even as a talisman: a new science was revealed, the dactyliomancia, so named, from two Greek words, signifying *ring* and *divination*. The performance of its mysteries was in itself so simple, that it was deemed expedient to add certain formulas, in order to make them more expressive. A ring was held suspended by a fine thread, over a round table, on the edge of which were placed counters engraved with the letters of the alphabet. The thread was shaken until the ring, touching the letters, had united as many as formed an answer to a question previously put. This operation was preceded and accompanied by certain ceremonies. The ring was consecrated with divers mysterious forms. The person who held it was arrayed in linen only; a circle was shaved round his head, and in his hand he held a branch of verveine. Before commencing the operation, the gods were appeased by prayer.

The Greeks had enchanted rings, in the construction of which the position of the celestial bodies was of the utmost importance.

The northern nations had also their magic rings. A very precious relic of the most remote Saxon antiquity was found, in 1841, on the borders of Rockingham Forest, in the parish of Cottingham, near Rockingham. This long-hidden witness of the past is a massive ring of pure gold, on which, in well-preserved Saxon characters, are two inscriptions. The outer one is as follows: "Guttu-Gutta-Mauros-Adros," and the inner: "Udro-Udros-Thebal." This ring is probably an *abraxis*, an amulet, or charm, worn as a preventive of evil.

Rings, with Runic inscriptions, have been found in Scandinavia as well as in the Saxon *barrows* in Great Britain. The Romans did not teach the fashion of finger-rings to the nations they conquered in Gaul or in Britain. They found them, at the period of their invasion, wearing this ornament on the third finger of the left hand.

Charmed rings have been believed in by many nations and in all ages, the only difference consisting in the ceremony used, and the name of the power supposed to impart the virtue. Whether consecrated by the invocation to a pagan divinity, or hallowed in the name of the God of Christians; whether endowed by eastern magi with supernatural influence, blessed by the sovereign pontiff, or hallowed by an English sovereign, the ring was still the chosen vehicle through which health, prosperity, and every other temporal good was conceded to man.

The kings of England had anciently a custom of hallowing rings on Good Friday, and these rings ensured the wearer against epilepsy. According to Hospinian, this custom took its rise from a ring which had long been preserved with great veneration in Westminster Abbey, and was supposed to be efficacious against cramp and epilepsy, when touched by those so afflicted. This ring, which had been brought from Jerusalem to King Edward the Confessor, was one which he himself had long before given privately in alms to a poor man for the love he bore to St. John the Evangelist. Tradition says, the saint himself brought the ring back to the king. These cramp rings, hallowed by the touch of royalty, continued to be dispensed by the sovereigns of England until the accession of the House of Hanover.

When William the Conqueror landed in England, he wore a ring hallowed by Pope Hildebrand.

Rienzi Gabrini, the last of Roman tribunes, obtained of one of the cardinals a hallowed ring, to which he attached implicit faith as a strengthener of his union with the Imperial city.

Talismanic rings continue to be worn in the East even now. In a work entitled *The Customs of the Mussulmans of India*, by Dr. Herklots, is given the following formula for the making of a ring, by the means of which princes may become subject to our will:—"Should any one desire to make princes and grandees subject and obedient to his will, he must have a silver ring made, with a small square tablet fixed upon it, upon which is to be engraved the number that the letters composing the *ism* represent, which in this case is 2·613. This number by itself, or added to that of its two demons, 286 and 112, and its genius 1·811, amounting in all to 4·822, must be formed into a magic square, of the *solacee* or *robaee* kind, and engraved. When the ring is thus finished, he is for a week to place it before him, and daily, in the morning and evening, to repeat the *ism* five thousand times, and blow on it. When the whole is concluded, he is to wear the ring on the little finger of the right hand.

CHAPTER VII.

RINGS.

The Sigillarius, or Seal-Ring.—Rings of Alexander, Sylla, Cæsar, Pompey, Augustus, and Mæcenas.—Seal Ring the Prerogative of the Wife or Eldest Daughter.—The Episcopal Ring.—The Annulus Piscatoris, or Fisherman's Ring.—The Annulus Sponsalium, or Nuptial Ring.—A Kiss, a Ring, and a Pair of Shoes.—Armenian Betrothals.—The Doge's Ring.—The Gimmal Ring.—The Mourning Ring.—Rings given at Weddings.—The Prodigal Philosopher.—Rings in the Middle Ages.—*Vie et Bagues Sauves.—Une Bague au Doigt.*—An Arab Saying —The Thumb Ring.—Poison in Rings.—Rings of Demosthenes and Hannibal.—Roman Lovers.—Rings as Souvenirs, Signals, Passports, Safeguards.—Devices in Rings —A Persian Custom.—Fashion for Rings under Henry III., Louis XVI., The Directoire.—Love's Telegraph.—Letter of Pope Innocent to King John.

" A contract of eternal bond of love,
Confirm'd by mutual joinder of your hands,
Attested by the holy close of lips,
Strengthen'd by interchangement of your rings."

" Sæpe velut gemmas ejus signumve probarem,
Per causam memini, me tetigisse manum."—Tibul. L. i. V. iv.

The *sigillarius*, or seal-ring, though not the richest in itself, was among the ancients the most important. Quintus Curtius states, that Alexander sealed the letters he sent into Europe with his own seal, and those he sent into Asia with that of Darius. When on his death-bed, he is supposed to have named his successor by drawing his signet ring from his finger, and giving it to Perdiccas.

The most celebrated engravers after Pyrgoteles, who lived in the age of Alexander, were Apollonides and Cromius, and also Dioscorides, who engraved a very good likeness of Augustus; an effigy subsequently used as a seal by the Roman emperors.

The dictator Sylla always used a seal representing the

betrayal of Jugurtha. Cæsar had on his the image of Venus; Pollio, that of Alexander; Pompey, that of a frog.

Upon Cæsar's arrival in Egypt, the murderers of Pompey, who brought the head of the vanquished hero to propitiate the conqueror, did not forget to add his ring also to the bloody present.

The Emperor Augustus had found among his mother's jewels two seal-rings exactly similar, representing a sphinx. During the civil wars, his friends, in his absence, used one of these rings to seal such letters and decrees as circumstances rendered necessary should be issued in his name: those whom they concerned were wont to say that these sphinxes did indeed convey enigmas. To avoid sarcasms, Augustus subsequently used a seal bearing the head of Alexander.

The seal of Mæcenas was an unwelcome apparition, denoting the levying of taxes.

In the civil law, seals were indispensable to establish the validity of documents, and were required from at least seven witnesses at the attestation of every will.

Among the Romans, the seal-ring belonged to the wife, and betokened her prerogative of having the charge of the valuables. As there were not then, as in modern times, locks and keys to every piece of furniture, precious articles, such as jewels, were kept in caskets sealed by the mistress of the house.

A widower, on his death-bed, delivered his seal-ring to his eldest daughter:—"*Annulum custodiæ causa majori natu filiæ tradidit.*"

As a symbol of spiritual alliance, and an insignia of clerical dignity, the ring dates back as far as the fourth century, in which we find it used in the ceremonies of the consecration of bishops. In 633, when the fourth Council of Toledo decreed that the ring should be restored to bishops who were reinstated after having been unjustly deposed; the decree merely confirmed a ceremony already ancient in the consecration. In the formula of blessing the episcopal ring, this ceremony is looked upon as the seal of faith, and a token of celestial protection. The ring is a symbol of the spiritual union subsisting between the bishop and the church. The same idea is conveyed in the words which the prelate who officiates at the consecration utters when he places the ring on the fourth finger of the new bishop.

Bishops were buried with a ring; and we are told by Matthew Paris, that Hubert, Archbishop of Canterbury, was made ready for the grave, clothed in his robes, with his face uncovered, his mitre on his head, gloves on his hands, and a ring on his finger, with all the other ornaments pertaining to his office.

Bishops formerly wore the ring on the forefinger of the right hand; but as they were obliged to remove it to the fourth finger during the performance of the holy rites, custom sanctioned its remaining there altogether.

The episcopal ring should be of gold, set with some rich gem, usually an amethyst, but not engraved, it having been so ordained by Pope Innocent III. This restriction has not, however, been strictly observed.

The bishops of the Greek Church wear no ring, this insignia being reserved for archbishops. In the Church of Rome, the right to the ring has extended from bishops and archbishops to cardinals; the latter, on receiving it, pay a certain sum—*pro jure annuli cardinalitii.*

The privilege of wearing a ring was subsequently extended to abbots.

The sovereign pontiff has two different seals: one, a large ring, the *annulus piscatoris*, the special seal of the popes, was already in use in the thirteenth century, and, as its name denotes, bears the effigy of Saint Peter drawing his nets. This ring serves for the apostolic briefs, and for private letters. It is used only by the pope himself, or in his presence and with his sanction. It is kept by the pope, or confided by him to the care of a member of the Sacred College.

The other seal, which is used for bulls, has the head of Saint Peter on the right, and that of Saint Paul on the left; a cross is between the two apostles. On the other side of this seal is the pope's name, and sometimes, though rarely, his arms. The seal for briefs is used with red wax; and that for bulls, with lead. At the decease of the reigning pontiff the seals are broken by the Cardinal Chamberlain, and new ones are offered by the city of Rome to the successor elected by the conclave.

The ring has played an important part in the investiture of the sovereign dignity.

Offa, king of the East-Angles, appointed his successor with the *ring*, the same one withal with which he had himself been invested when promoted to the throne.

Richard II., when compelled to resign the crown of England to Henry of Lancaster, did so by transferring to him *his ring*. A paper was put into the hands of the humbled monarch, from which he read an acknowledgment of his unfitness to reign, and of the justice of his deposition: also that he freely absolved his subjects from their allegiance, and swore, by the Holy Gospels, never to act in opposition to his surrender: adding, that if it were left wholly to him to name his successor, it should be Henry of Lancaster, to whom he gave his ring.

The coronation ring of the kings of England is of plain gold, with a large violet table ruby, whereon a plain cross, or cross of St. George, is curiously enchased.

The Queen's ring is likewise of gold, with a large table ruby, and sixteen small brilliants around the ring.

In conformity with the ancient usage recorded in Scripture, the primitive Christian Church early adopted the ceremony of the *Sponsalium Annulus*, or ring of affiance, in marriage, as a symbol of the authority which the husband gave the wife over his household, and over the earthly goods with which he endowed her.

Leobard, the celebrated Saint of Tours, in the sixth century, being persuaded in his youth to marry, gave his betrothed a ring, a kiss, and a pair of shoes.

In some of the Pompeian pictures, brought to light after an inhumation of one thousand seven hundred and sixty years, the female figures are represented wearing *intagli* and *camii*, set in rings placed on the wedding finger.

The Roman custom of wearing the ring on the fourth finger of the left hand was confirmed by the Christian Church. In the ancient marriage ritual, the husband placed the ring first on the first joint of the bride's thumb, saying: "In the name of the Father;" he then removed it to the forefinger, with the words, "In the name of the Son;" then to the middle finger, adding, "And of the Holy Ghost;" finally, the ring was left on the fourth finger, with the word "Amen!"

It seems that about a century ago, it was the custom to wear the marriage-ring on the thumb, although at the nuptial ceremony it was placed on the fourth finger (*vide British Apollo*). It was formerly believed that to that finger an artery of the heart reached, and hence its being adopted for the wedding-ring.

Strutt tells us, he finds no mention of the marriage-ring in the Saxon era, except in the *Polychronicon*, translated by Trevisa, who tells a story of a young man at Rome, (in the time of Edward the Confessor,) who being at play on his wedding-day, "did place his spousing rynge on the fynger of an ymage" of Venus, and could by no means get it off; at night the statue interposes between him and his bride, claiming him as her spouse. This story has been put into verse by German and English authors, and is a proof of the binding power formerly, and even now, attributed to this little circlet. What woman's heart is not assailed by superstitious fears, if she lose her wedding-ring?

Among the Armenians, children are betrothed from their earliest youth, sometimes when only three or four years old, sometimes as soon as born. When the mothers on both sides have agreed to marry their son and daughter, they propose the union to their husbands, who always sanction the choice of the wives. The mother of the boy then goes to the friends of the girl with two old women and a priest, and presents to the infant maiden a ring from the future bridegroom. The boy is then brought, and the priest reads a portion of the Scripture and blesses the parties. The parents of the girl make the priest a present in accordance with their means, refreshments are partaken of by the company, and this constitutes the ceremony of the betrothals. Should the betrothals take place during the infancy of the contracting parties, and even should twenty years elapse before the boy can claim his bride, he must every year, from the day he gives the ring, send his mistress, at Easter, a new dress with all the belongings thereunto, in accordance with her station in life.

Among the most famous espousals were those celebrated by the Doge of Venice with the Adriatic, which dates from the middle of the twelfth century. This singular ceremony is said to have originated with the gift of a gold ring presented to Sebastiano Ziani, on his defeat of the ships of the Emperor Barbarossa, by Pope Alexander.

The pope, on the return of the victors to Lido, hastened to receive them; and as soon as Ziani touched the shore, he placed on his hand a ring of gold, (the antique Roman badge of power,) saying: "Take this ring, and with it take, on my authority, the sea as your subject. Every year, on the return of this happy day, you and your successors shall make known to all posterity that the right of conquest has subjugated the Adriatic to Venice, as a spouse to her husband." And annually, on Ascension Day, through the long course of six hundred years, the magnificent galley, *The Bucentaur*, bore the doge to the shores of Lido, near the mouth of the harbour. Here, letting a ring fall into the bosom of his bride, the bridegroom uttered the "*Desponsamus te mare, in signum veri perpetuique dominii.*"—"We wed thee with this ring in token of our true and perpetual sovereignty." The boast was presuming too much on the duration of mortal dominion. The union of the doges with the Adriatic was as subject to be severed as that of common mortals, and the fall of the republic witnessed the divorce of this aquatic marriage, and the end of the superb annual pageant.

The gimmal ring is of comparatively modern date, and of French origin. It is composed of twin or double hoops, which, though each is twisted, fit so exactly one into the other that, when united, they form but one circlet. Each hoop is usually surmounted by a hand raised somewhat above the circle, and, when the hoops are brought together, each hand clasps its fellow. One hoop was sometimes of gold and the other of silver; they were then divided, one being worn by the lover and the other by his mistress. This kind of ring is alluded to in Beaumont and Fletcher's *Beggar's Bush*:—

> "*Hub.* Sure I should know that gimmal.
> *Jac.* It's certain he—I had forgot my ring, too."

These little circlets,—emblems of eternity in their shape,—were symbols of love strong in death as well as in life; memorials which the departed left to their surviving friends. The custom of mourning rings is an ancient one. Shakspeare bequeaths to his *fellows*, John Henninge, Richard Burbage, and Henry Condell, "twenty-six shillings eight-pence a-piece to buy them rings."

Rings were in olden times given away to the attendants

on the day of marriage. We find in Wood's *Athenæ Oxonienses*, that the once famous philosopher, Kelley, who was prodigal beyond all bounds of wisdom, "did give away in gold wire rings, (or rings twisted with three gold wires,) at the marriage of one of his maide-servants, to the value of *four thousand pounds*." This took place at Trebona, in 1589.

In the life of the infamous Judge Jeffreys we read, that on the 17th February, 1668, he was called serjeant, on which occasion he gave rings with the motto, "The King from God, and the law from the King."

The ring in the middle-ages was a badge of knighthood, as it had been with the Roman equites, and was given as a pledge for the fulfilment of promises. The word *bagues*, rings, was, in French, in certain cases, synonymous with baggage, personal effects. Hence the expression employed in capitulations, *sortir vie et bagues sauves*, literally to depart with life and rings safe. It probably originated in the importance attached to this insignia: it is frequently used in the history of the fourteenth and fifteenth centuries.

When Rouen was retaken by Charles VII., the English were permitted to retire with their effects. They asked leave to go forth free with their rings, which was granted, "on condition," said the king, "that on the road they should take nothing without paying for it." They replied, that they had not wherewith; the king then ordered that a hundred francs should be given to them*.

Robert de Launoy, who commanded in Hesdin, in 1477, capitulated, and was allowed to retire with his garrison *vies et bagues sauves*.

In terms of jurisprudence *bagues et joyaux*, rings and

* "*Ils démandèrent à sortir vie et bagues sauves. 'A condition,' dit le roi, 'que sur la route, ils ne prendront rien sans payer.' 'Nous n'avons pas de quoi,' répondirent ils; le roi leur fit donner cent francs.*"—M. DE BARANTE: *Histoire des Ducs de Bourgogne.*

jewels, formerly signified all the personal valuables of a woman, even her costly apparel, or the money presented in their stead by the marriage contract. The expression is no longer used.

The French expression, *une bague au doigt*, a ring on the finger, means a sinecure, any post that brings a handsome profit without requiring much exertion.

Among the Arabs, *to put on a ring* means to get married.

The fashion of wearing a ring on the thumb is very ancient in England. In Chaucer's *Squior's Tale*, it is said of the rider of the brazen horse, who advanced into the hall of Cambustan, that—

"——— Upon his *thomb* he had of gold a ring."

When the tomb of Bede was opened on May 27, 1831, a large thumb-ring of iron, covered with a thick coating of gold, was discovered in the place which the right hand occupied before it had fallen into dust. The device is a *cinque-foil.*

In the reign of Queen Elizabeth, grave persons, such as aldermen, used a plain, broad, gold ring upon their thumbs.

An alderman's thumb-ring is mentioned (says Stevens) by Brome, in the *Antipodes*, in 1638, which was acted with great applause at Salisbury Court, in Fleet Street; also in the *Northern Lass*, which was held at the Globe and Black Friars, in 1603, viz.: "A good man in the city wears nothing rich about him but the gout or a *thumb-ring*." Again, in *Wit in a Constable*, 1660, viz.: "No more wit than the rest of the Bench; what lies in his *thumb-ring*."

Rings have been used in ancient and modern times as secret and safe repositories for deadly poisons. Orators and warriors have had recourse to these means. Demosthenes put an end to his life either with a poisoned feather, or with the poison he carried in his ring. The greatest general of his age, being betrayed by his host, destroyed himself with the poison contained in his ring. What was the substance thus used by either the Greek or the Carthaginian is not known. Modern chemistry has a variety of preparations that might be introduced into a ring, and, under certain circumstances, destroy life. The Carthaginians were not likely to be acquainted with prussic acid. Lybia, however, then furnished them with deadly poisons, and it was probably one of these that the provident hero, who knew his countrymen so well, had kept in his ring.

Under the third consulate of Pompey, two thousand pounds weight of gold, which had been deposited in the throne of Jupiter Capitolinus by Camillus, were stolen. The officer in charge having been arrested, crushed the setting of his ring between his teeth, and died on the spot, thus annihilating all clue to the discovery of the thief, by the destruction of the sole witness.

Roman lovers seem to have made use of the same little subterfuges as the beaux of our own day. Tibullus confesses, that he found means to press the hand of his mistress, under pretence of admiring the gems that adorned her fingers.

Rings have been more frequently used as signals, passports, tokens of recognition, etc., than any other article. The ring has been as instrumental in saving as in taking lives. In the life of Mary, Queen of Scots, we find the queen's ring rescuing the condemned one, even on the scaffold. In the linked histories of that queen and her rival Elizabeth, we have innumerable instances of rings going backwards and forwards, and playing a very important part in the great drama of their lives, as souvenirs, love-tokens, warnings, passes, etc. The death of Queen Mary of England was signified to her expectant successor by the envoi of a black enamelled ring, which the queen wore day and night. The ring was to be removed from the finger of the corpse as soon as life was extinct, by one of the ladies of the bedchamber, and dispatched to the princess Elizabeth, through Sir Nicholas Throckmorton. The incident is given in the metrical chronicle of the life of that worthy.

> "She said, since nought exceedeth woman's fears
> Who still do dread some baits of subtlety,
> Sir Nicholas know a ring my sister wears,
> Enamelled black, a pledge of loyalty,
> The which the king of Spain in spousals gave,
> If ought fall out amiss, 'tis that I crave."

The prudent Elizabeth little thought her own death would be similarly communicated to *her* successor. When that event occurred, one of her ladies drew from her royal mistress' lifeless hand a turkois ring, and threw it out of a window to Sir John Harrington, as had been preconcerted between them. By him the token was conveyed with all speed to James VI. of Scotland.

A ring was the indirect cause of the deaths of both Mary of Scotland and Elizabeth of England. A diamond ring, sent by the English queen, as a warrant of her friendship and good faith to poor Mary, led the latter to trust herself into the hands of her implacable foe, to perish on the scaffold at last*. The ring sent by Essex to his royal

* This memorable ring is described by Aubrey, the great antiquary, to have been a delicate piece of mechanism, consisting of separate joints, which, when united, formed the quaint device of two right hands supporting a heart between them. This heart was composed of two separate diamonds, held together by a central spring, which, when opened, would allow either of the halves to be detached. The circumstance of the ring is further verified beyond dispute by Mary herself, in a subsequent letter to Elizabeth, in which she bitterly reproaches her with her perfidious conduct. "After I had escaped from Lochleven," she says, "and was nearly taken in battle by my rebellious subjects, I sent you, by a trusty messenger, the diamond you have given me as a token of affection, and demanded your assistance. I believed that the jewel which I had received as a pledge of your friendship, would

mistress, the withholding of which caused his death on the scaffold, and hers in agonies of grief, was the instrument through which retributive justice was dealt on her who had treacherously broken the word given on a pledge of the same kind. Elizabeth, when in a tender mood, had bestowed this ring upon Essex, as a warranty, that in his utmost need, he should find favor from her. After his condemnation to death, the fallen favorite had sent the love-token to her who could alone save him. The ring, thrown by the prisoner from a window in the Tower, to a boy, with a hurried verbal message, was taken by mistake to the sister of the lady who was to have conveyed it to the queen. This sister was the wife of the Earl of Nottingham, the worst foe of the unhappy Essex; and she, persuaded by her husband, kept it. Elizabeth, exasperated by what she deemed the perverse obstinacy and pride of him she still loved, signed his death-warrant. When the Countess, on her death-bed, revealed to Elizabeth this touching appeal made by Essex for his life, it is recorded that Elizabeth, in her rage, shook the dying woman, exclaiming:—"God may forgive you, I never will!" The queen from that moment refused all food, and gave herself up to the deepest melancholy, dying shortly afterwards.

This was the last of the little blood-stained circlets that acted so busy a part during that eventful reign. Many, if not all, of these mute witnesses of the tortures of princely hearts, may yet be in existence; and, for aught we know, be again emissaries of weal or woe.

When the Duchess of Savoie was held a prisoner by the irritated Duke of Burgundy, Charles the Bold, she found

remind you, that when you gave it to me, I was not only flattered with great promises of assistance from you; but you bound yourself, on your royal word, to advance over your border to my succour, and to come in person to meet me, and that, if I made a journey into your realm, I might confide in your honor."—Gilbert Stuart, vol. ii., page 232.

means to send her secretary to solicit the aid of Louis XI. As she was prevented from writing, the only credentials she could give her emissary was the ring the king had given her on the occasion of her marriage. This passport would have proved all sufficient, but that unfortunately the bearer, when he presented himself to the king, wore the cross of St. André. The suspicious monarch, who was inclined to judge of all men by his own deceitful heart, immediately ordered the man to be arrested, supposing him to be a spy of the Duke of Burgundy, and that he had stolen his sister's ring. The hapless secretary was in imminent danger of being hung, when he was saved by the timely arrival of the Lord of Rivarola, who was sent by the duchess, urging the king to hasten to her rescue.

The importance thus given to the ring is alluded to by Shakspeare :—

"If entreaties
Will render you no remedy, this ring
Deliver them, and your appeal to us
There make before them."

Henry VIII., v. 1.

"Look there, my lords:
By virtue of that ring, I take my cause
Out of the gripes of cruel men, and give it
To a most noble judge, the king, my master.
"This is the king's ring.
"'Tis no counterfeit.
"'Tis the right ring, by Heaven!"

Idem., v. 2.

The old fashion of devices, mottos, or, as they were formerly called, *posies*, on rings, has been revived within a year, and the jewellers' windows once more present golden and enamelled hoops inscribed with some word or maxim. Shakspeare has more than one allusion to these love-gifts.

Gratiano, complaining of the quarrel Nerissa fastens upon him, says, it is—

> "About a hoop of gold, a paltry ring
> That she did give me; whose poesy was
> For all the world, like cutler's poetry
> Upon a knife, *Love me and leave me not!*"

And the melancholy Jaques, in *As You Like It,* says to Orlando,

> "You are full of pretty answers;
> Have you not been acquainted with
> Goldsmiths' wives, and conned them out of rings?"

These devices were sometimes expressed by precious stones;—for instance, the word *regret* was spelt with a ruby, an emerald, a garnet, a ruby, and a topaz; the name *Adèle* was spelt with an amethyst, a diamond, an emerald, a lapis-lazuli, and another emerald.

According to Tavernier, the Persian jewellers never make gold rings; the religion of the Persians forbidding the wearing of any article of that metal during prayers, it would be too troublesome to take them off every time they perform their devotions. The gems mounted in gold rings, sold by Tavernier to the king, were reset in silver by native workmen.

Finger-rings have never ceased to be worn, though fashion at times has limited or increased their number. In the reign of Henry III., of France, three were worn on the left hand; one, on the second finger; one, on the third, and one, on the fourth.

Never, since the days of the Romans, had rings been worn in such numbers as they were in France at the close of the last century. The ring and snuff-box mania was carried to such an extreme, that the minister of a German court had a ring and a snuff-box for every day in the year. Snuff-boxes were made light or heavy, according to the season for which they were intended. A contemporary writer tells us, that, in 1788, enormous rings were worn. The hand of a woman presented a collection of rings, and had these been antiques, might have been taken for a sample of a cabinet of engraved gems. He adds, that "the nuptial ring is now unnoticed on the fingers of women; wide and profane rings altogether conceal this warrant of their faith*."

Rings have not been worn on the fingers or in the ears only; the ankles and the toes have been similarly adorned; the chin, cheeks, and nostrils, have been disfigured by them. Many eastern nations still wear rings on the toes. The kings of Pegu wear rings set with precious stones on each toe. Sometimes the wing case of the diamond-beetle, or the whole insect, is mounted like a gem on rings, and has a very pretty effect.

The Turkish ladies, who spend their lives reclining on cushions, wear rings on their toes as well as their fingers.

Many of the Cingalese women wear toe-rings.

But we need not go to the East to find the toe-ring. At a no very distant period, it reigned in the capital of civilized fashion. When the belles who flourished under the short-lived reign of associated kings, yclept the Directoire, adopted, without regard for difference of climate, what they deemed classical costumes, they did not omit toe-rings; and the fashionables of the day made their appearance in the gardens of the Tuilleries, with unstock-

* "On porte actuellement des bagues énormes, et la main d'un Turcaret n'est plus chose rare; les hommes font la belle main.

"La main d'une femme est un baguier, et si ses bagues étaient des antiques, elles offriraient un échantillon d'un cabinet de pierres gravées: aussi l'anneau nuptial est il inapperçu chez nos femmes; des bagues larges et profanes étouffent ce gage de leur fidélité."—MERCIER: *Tableau de Paris*. Vol. ii.

inged feet and sandals, in which their toes, ornamented with gems, were fully displayed.

Among the Romans, women of a certain class, who affected independence of all received customs and fashions, and were comprehended under the denomination of *libertinæ*, wore above their ancles, gold rings, set with precious stones.

Rings have been made to contain all sorts of affectionate *souvenirs* and diminutive toys: hair, portraits, watches, &c. But one of the most singular articles of *bijouterie* is mentioned in Thiebault's *Original Anecdotes of Frederic II.* M. de Guines, ambassador of France at Berlin, had greatly mortified the Prussian nobles, and especially the other foreign ministers, by the ostentatious pomp which he displayed. Those whose limited means he thus eclipsed longed for some opportunity to wound the vanity of the proud man who daily humbled theirs, and excited their envy. At this crisis, a Russian ambassador, who was returning home to present at his own court his newly-married bride, stopped on his way at Berlin. Prince Dolgorouki, the Russian ambassador there, did the honors of the Russian court to his countryman, and gave him and his wife a dinner, to which were invited all the corps diplomatic. M. de Guines was seated next to the bride. The lady, who had been initiated into all the court gossips, had enlisted under the banner of the malcontents, and taken upon herself the task of vexing the magnificent Frenchman. She had placed upon her finger a ring, of very exquisite and very curious workmanship, to which she called the attention of her neighbour during the course of the dinner. As he stooped to examine the jewel, the wearer pressed a spring concealed on the side of the ring within her hand, and jerked a small quantity of water into the eyes of the ambassador. The ring contained a syringe. The minister wiped his face, jested good-humoredly on the diminutive little instrument, and thought no more of it.

But his fair enemy had not yet accomplished her purpose of mortifying the ambassador. Having refilled the squirt unperceived by him, she called his attention to herself, and again discharged the water in his face. M. de Guines looked neither angry nor abashed, but, in a serious tone of friendly advice, said to his foolish aggressor,—"Madame, this kind of jest excites laughter the first time; when repeated, it may be excused, especially if proceeding from a lady, as an act of youthful levity; but the third time it would be looked upon as an insult, and you would instantly receive in exchange the glass of water you see before me: of this, Madame, I have the honor to give you notice." Thinking he would not dare to execute his threat, the lady once more filled and emptied the little water-spout at the expense of M. de Guines, who instantly acknowledged and repaid it with the contents of his glass, calmly adding,—"I warned you, Madame." The husband took the wisest course, declaring the ambassador was perfectly justified in thus punishing his wife's unjustifiable rudeness. The lady changed her dress, and the guests were requested to keep silence on the affair: an injunction obeyed as is usual in such cases.

A ring in all probability saved the Emperor Charles V. from the most critical position in which he was ever placed. Having requested permission of Francis I. to pass through France, the sooner to reach his Flemish dominions, where his presence was urgently required, the rival so lately his prisoner not only granted the request, but gave him a most brilliant reception. Some of the French king's counsellors thought this generous conduct towards so crafty a foe was quixotic in the extreme, and that Charles should be detained until he had cancelled some of the hard conditions to which he had compelled Francis to subscribe to purchase his release. Among those who strongly advocated the policy of detaining the imperial guest, was the king's fair friend, the Duchesse d'Estampes. Charles, who was informed of the dangerous weight thrown in the scale against him, resolved to win over the influential counsellor. One day, as he was washing his hands before dinner, he dropped a diamond ring of great value, which the duchess picked up and presented to him. "Nay, Madame," said the Emperor, gallantly, "it is in too fair a hand for me to take it back." No more was said to Francis on the folly of keeping knightly faith with one who had shown him so little mercy; and the Emperor was permitted to pursue his way in all freedom, rejoicing that his good luck had been greater than his prudence.

Rings in modern times have been made, in some countries, love's telegraph. If a gentleman wants a wife, he wears a ring on the first finger of the left hand; if he be engaged, he wears it on the second finger; if married, on the third; and on the fourth, if he never intends to be married. When a lady is not engaged, she wears a hoop or diamond on her first finger; if engaged, on her second; if married, on the third; and on the fourth, if she intends to die a maid. As no rules are given for widows, it is presumed that the ornamenting of the right hand and the little finger of the left, is exclusively their prerogative. This English fashion is, perhaps, too open a proclamation of intentions to suit such as do not choose to own themselves as mortgaged property.

The following letter, witten by Pope Innocent, to John, king of England, showing the symbolical use then made of rings and precious stones, will prove an appropriate conclusion to the account here given, of an ornament universally admired in all ages.

Pope Innocent, to John, King of England.

"Among the riches that mortals prize as the most valuable, and desire with the greatest earnestness, it is our opinion that pure gold and precious stones hold the first rank. Though we are persuaded that your royal excellence has no want of such things, we have thought proper to send you, as a mark of our good-will, four rings, set with stones. We beg the favour, you would consider the mystery contained in their form, their matter, their number, and their colour, rather than their value. Their roundness denoting eternity, which has neither beginning nor end, ought to induce you to tend, without ceasing, from earthly things to heavenly, and from things temporal to things eternal. The number four, which is a square, signifies firmness of mind, not to be shaken by adversity, nor elevated by prosperity, but always continuing in the same state. This is a perfection to which yours will not fail to arrive, when it shall be adorned with the four cardinal virtues, justice, fortitude, prudence, and temperance. The first will be of service in judgments; the second, in adversity; the third, in dubious cases; the fourth, in prosperity. By the gold is signified wisdom; as gold is the most precious of metals, wisdom is, of all endowments, the most excellent, as the prophet witnesses in these words: 'The spirit of wisdom shall rest upon him.' And, indeed, there is nothing more requisite in a sovereign. Accordingly, Solomon, that pacific king, only asked of God wisdom, to make him to well govern his people. The green colour of the emerald denotes faith; the clearness of the sapphire, hope; the redness of the ruby, charity; and the colour of the topaz, good works—concerning which our Saviour said, 'Let your light so shine before men, that they may see your good works.' In the emerald, therefore, you have what you are to believe; in the sapphire, what you are to hope; in the ruby, what you are to love; and, in the opal, what you are to practise; to the end you may proceed from virtue to virtue, till you come to the vision of the God of gods, in Sion."

CHAPTER VIII.

PAWNED JEWELS.

Jewels a ready Resource.—Many an unsuspected *Parure* acquainted with *My Uncle*.—Jewels a safe Investment during the Middle Ages.—Ancient Romans *au fait* in the Mysteries of Pawning.—Vitellius pledges his Mother's Pearl. —The Sand-filled Coffers of the Cid.—Henry III. pledges the Virgin, and Edward III. his best Friend.—The Black Prince, Henry V., Henry VI., and Richard II.—Jewels of the Great continually travelling back and forth. Jewels of the Duke of Burgundy pawned.—Poverty of the King of France, and Rapacity of his Nobles.—Jewels of the Dukes of Orleans pledged, of Elizabeth of York, Henry VIII., Anne Boleyn, James VI., Henrietta Maria, and Mary Beatrice of Modena.—Napoleon I. and the Regent.—Annual Report of the Mont-de-Piété.—Diamonds of Mademoiselle * * * * * ten years at the Mont-de-Piété.

"O, my prophetic soul! My uncle!"

No description of property offers as ready, as instantaneous, a resource, in moments of pecuniary difficulty, as jewels; and of all jewels, diamonds. No sacrifice of cherished souvenirs, of family heir-looms; no selling, for a mere song, articles of inestimable value, is imposed now by the stern necessity of the hour. A short walk in the dusk of evening, a few minutes' interview with the accommodating employé, and the borrower returns with a heavier purse, a lighter heart, a more elastic step; leaving his valuables in safe keeping. He has, moreover, the consciousness that the secret of his embarrassments will be as safe as his property, and that he can redeem the latter whenever he has the sum to repay that which he has just obtained. Could the brilliant *parures* that sparkle on heads and bosoms at the court balls relate their wanderings, very many of those least likely to be suspected would be found to have made a longer or shorter sojourn, on one or more occasions, at the residence of that convenient and universal, but unacknowledged, relative of the Rue des Blancs, Manteaux. Many an unlucky *coup-de-bourse*,—the debt of honor of the gambler of the nineteenth century,—has sent the glittering trinkets into temporary retirement, whence they have again issued in undimmed glory, leaving the tell-tale duplicate behind them.

We are rather disposed to exclaim against the extravagance of our ancestors, without considering that in the times that preceded the invention of those ingenious improvements upon their financial arrangements,—consols, and reduced three per cents,—and when the lending out money for hire was considered a disgraceful transaction, and almost confined to the Jews, plate and jewels were the most convenient, if not the only investments of spare funds, being convertible into cash, upon pledge, or by sale, at will.

Unfortunately, while immense sums were lavished on jewels and plate, other expenses were proportionately large, other fancies kept pace with the taste for jewelry, and were indulged to as great an extent; but with this difference, that of these nothing remained after the gratification of the moment; fine clothes were worn out, and banquets were consumed, while the money expended in valuable gems remained, in another form, but one quite as available in case of need.

There have been few sovereigns from the ninth to the nineteenth century who have not, at some period of their lives, had occasion to pledge their jewels. When invasions, foreign or domestic wars, revolutions, famines, or miscalculations, have beggared the public exchequer, and emptied the privy purse, jewels have been found a portable capital, and a ready resource. The Jews and the Lombard money-changers, high dignitaries of the church, and rich citizens, nay, even proud cities, such as Venice, London, and Genoa, were formerly the pawnbrokers of princes.

Although the Lombard money-changers of the middle-ages, the government-established Mont-de-Piété of modern Gaul, and the convenient *three balls* of our English neighbours, were unknown to the ancients, they appear to have had recourse to similar resources when their finances were at a low ebb. The Emperor Augustus, in the beginning of his reign, created a fund, from the confiscated property of criminals, for lending on pledges which were double the value of the sum required, but without interest. Tiberius followed this example, and lent to the poor on landed security. Were we acquainted with the slang phrases of the fast young Romans of Cæsar's day, we should probably find them as elegant and expressive as *chez ma tante, mettre en plan, mettre au clou;* the *pop-shop, going to your uncle's, raising the wind,* and *putting up the spout,* of the present generation.

When Vitellius obtained of the Emperor Galba the command of the German legions, he was so poor that, in order to defray the expenses of his journey, he was not only obliged to leave his wife and children in a small room and rent his house during the remainder of the year, but he also took from his mother a large pearl she wore in her ear and pledged it. "Atque ex aure matris detractum unionem pigneraverit ad itineris impensas*."

If we consult the annals of the eleventh century, we find the famous Cid Campeador, from whom the royal race of Spain hold it an honor to trace their descent, pledging his *locked* treasure-coffers to the sons of Israel, who, to their praise be it recorded, more confident in the word of a Christian than their descendants, did not insist on seeing the contents! Had some unlucky Moorish scimitar cut short his glorious career before he had redeemed the pledges, an examination of the apocryphal jewels and plate would have somewhat tarnished his fame.

* Suetonius.

The extremes to which the uxorious Henry III. of England and his queen were often reduced, may be imagined when we find, that, in the twenty-seventh year of his reign, he issued an order, directing that the most valuable image he possessed of the Virgin Mary should be pledged, to obtain a sum of money required for the payment of the officers of the chapel-royal at Windsor. It was especially stipulated that the hallowed pledge should be deposited in a decent place.

During the long reign of Edward III. the crown-jewels were seldom out of pawn, and *magnam coronam Angliæ*, his imperial crown, was pledged three several times,—once abroad and twice to his banker, Sir John Wosenham, in whose custody it remained no less than eight years. In the commencement of his long war, he was obliged to pawn his queen's crown, at Cologne, for two thousand five hundred pounds, in the year 1339. Soon after the people submitted to a tax, not *on* wool, but *of* wool, and subscribed thirty thousand packs of that commodity, which, being sent down the Rhine to Cologne, redeemed Philippa's best crown from thraldom.

The following year, the mighty Edward and his queen were literally in a state of bankruptcy. She had given up her crown and all the jewels she possessed, to be pawned to the Flemish merchants. But the necessities of the royal pair increasing, the English monarch actually pawned the person of his valiant kinsman, the Earl of Derby, who willingly surrendered himself a captive, while Edward stole away with his queen and her infant to Zealand.

The next year, Edward obtained supplies of his parliament by declaring, that if he was not able to redeem his honor and his cousin, the Earl of Derby, he would return to Flanders and surrender himself to his creditors. In answer to this appeal, the commons granted the fleece of the ninth sheep and the ninth lamb throughout England.

English coin seems to have been as scarce with the subjects, as with the royal master and mistress.

The Earl of Derby had been detained in prison by Matthew Concannen, and partners of the firm of the Leopard.

The Black Prince, as Walsingham informs us, was constrained to pawn his plate.

Henry V. sold or pawned all the valuables he possessed, among which were the silver tables and stools he had from Spain, to raise funds for his French expedition. The magnificent crown, called the Great Harry, was pawned. This sovereign beggared his own exchequer, rendered the crown bankrupt, and deluged the continent in blood, in the vain attempt to unite France and England under one sceptre.

Still poorer was the next sovereign. His bride, Margaret of Anjou, began her career as an English queen rather inauspiciously. When passing through Rouen, on her way to marry Henry, she was so short of money that she pawned vessels of *mock* silver to raise funds to continue her journey. In the meanwhile, her royal bridegroom was pawning all his jewels and household plate, to provide equipages and other requisites for her reception. Henry VI. pawned one of the crown jewels, called the "rich collar," in three several pieces, to different persons. He was never able to redeem it.

When Richard II. married Anne of Bohemia, in 1381, the jewels of the duchy of Aquitaine, the floriated coronet, and many brooches and clasps in the forms of animals, were pawned to the Londoners, in order to procure money for the expenses of the bridal. Rymer mentions, that in the ninth year of his reign, Richard pawned certain jewels *à la guise de cerfs blancs**.

* The badge of the *white hart* was assumed by Richard II., and worn by all his courtiers and adherents, both male and female,

During the whole of the fifteenth century, the jewels of the great seem to have been travelling back and forth between their owners and the money-lenders. Sometimes the persons of princes and peers were resplendent with *orfévrerie* and precious stones; their sideboards groaned under the load of gem-studded gold plate; then, again, all these riches were packed off to raise funds to sustain private broils or civil wars.

Princes and lords, on the French side of the channel, were neither wiser nor more thrifty than their English cousins. When Philip the Bold was preparing for his nuptials with the heiress of Flanders, in 1369, the duke, that he might appear with the magnificence he deemed necessary on such an occasion, procured all the pearls, diamonds, gems, and jewels of all sorts that could be purchased. Enguerrand, Lord de Coucy, alone sold to him jewels to the amount of *eleven million* francs. But the duke had been so princely in all things, so lavish in his own expenditure, and so munificent in his gifts, that four days after the nuptials, in order to return home, he was obliged to raise money, by pawning to three citizens of Bruges the jewels he had left: notwithstanding which he gave yet another splendid banquet to the chief men of the town, before he departed.

When the same duke visited the pope at Avignon, he kept such state, and made such rich gifts to the pontiff and the cardinals, that to defray the expense of his journey home, he was obliged to leave his jewels in pawn with a Lombard, for twenty thousand francs.

Everywhere we find the same lavish expense and constant penury. The Duke de Berry finds it impossible to hold out in his town of Bourges, besieged by the king and the Duke of Burgundy; he cannot pay his men-at-arms,

either embroidered on their dresses, or suspended by chains or collars round their necks.

though he has sold or pawned all his plate and jewels. A few days after, at an interview between the two dukes, in this same town of Bourges, the Duke of Berry makes a splendid appearance. "Both dukes were in armour. The Duke de Berry, who was over seventy years of age, was of a firm and noble countenance; the raised vizor of his helmet was adorned with precious gems; he wore the white scarf, embroidered with *marguerites;* a poniard at his belt, and a battle-axe in his hand, completed his martial costume*."

In a remonstrance, addressed by the university and municipality of Paris, to the king, Charles VI., on the disorderly and thievish manner in which his finances were administered by his treasurers, one of these is mentioned as having robbed him to such an extent, that "he was possessed of innumerable rubies, sapphires, and other gems; also magnificent garments; he kept great state; and in his house all the dishes, basins, ewers, cups, and goblets, were of silver." While the officers of the king's household were living in such splendour, their hapless master never had a *dénier*, even of the ten gold crowns daily put into the hands of Maurice de Reuilly, his cofferer, as pocket-money for the sovereign. When money was needed for war or other necessities, none was to be found. "Then," continued the remonstrance, "the merchants, who sell money, and have acquired it by usury and rapine, are sought; your majesty's jewels and plate are given to them in pawn, and you pay fifteen thousand francs to obtain ten thousand."

The princes and lords taken at Azincourt, pawned or sold all their plate and jewels to pay their ransoms. In 1417, Charles, Duke of Orleans, sold his plate to ransom his brother John, Count of Angoulême, prisoner with himself in England. In 1436, the same prince ordered a golden cross and a ruby to be sold at Bruges, the money

* *Histoire des Ducs de Bourgogne*, par M. de Barante.

the jewels brought being put into the hands of Dunois to be used, "for the good of his (the duke's) affairs." Some thirteen years later, however, the private affairs of Dunois himself must have been in a flourishing condition, for we find him, on the occasion of the entrance of Charles VII., in Paris, wearing a sword in a gold scabbard, garnished with diamonds and rubies to the amount of more than fifteen thousand crowns.

The liberal Duke of Burgundy, John the Fearless, son of Philip the Bold, was often, notwithstanding his ample means, reduced to pawn his old plate and jewels when he chose to purchase new. Among the jewels he pawned, are recorded: "A clasp of a hart, of a device of King Richard, garnished with XXII. large pearls, II. square balaxes, II. sapphires, and I. ruby; also, I. large square pointed dyamont in a gold setting, the which dyamont is of the size of a filbert nut; a gold clasp with a bear enamelled whyte, with two dyamonts about the throat thereof, a ruby and a large pearl pendant, and another ruby in front of sayd bear; also another gold clasp garnished with three pearls, a long ruby in the center, and a square pointed dyamont; on the top of the aforesaid clasp, are two flowers, the one whyte, and the other of gold; also, a thick enamelled gold crozier, garnished with six large round pearls, weighing about three carats each," &c.

During the fifteenth century, France, continually desolated by foreign wars and domestic broils, invaded by a pitiless stranger, and torn by her own children,—with one sovereign lingering a prisoner in England, a second helplessly insane during thirty years, and a third occupied all his life either in reconquering his kingdom from an invader, or from his nobles,—was too impoverished to permit of her king equalling in extravagance the great vassals who had despoiled him. In 1422, Charles VII. pawned his large diamond called the Mirror, in order to obtain the funds necessary to enable him to make the customary New-

Year's presents to his courtiers, who were certainly better able to give than he was. In 1435, Charles again borrowed, of the Bishop of Paris, two hundred *saluts* of gold, on a golden tablet representing the Trinity and Saint Margaret, garnished with a *very large* pearl and two large sapphires.

Notwithstanding the penury of the crown, Charles's fair friend, Agnes Sorel, the Lady of Beauty, attended the tournament given in honor of the marriage of Margaret of Anjou and Henry VI., in the dress of an amazon; her fanciful armour, and the caparison of her horse blazing with jewels. She, whose chivalrous advice had contributed largely to rescue the kingdom from a foreign tyrant, had, certes, a better right to such costly ornaments than those who had repeatedly sold king and country to procure them.

In 1495, the Duke of Orleans, subsequently Louis XII., pawned jewels at Lyons to the amount of five hundred and fifty gold crowns. So deeply in debt was Elizabeth of York, during this same year, that her consort, Henry VII., found it necessary, after she had pawned her plate for five hundred pounds, to lend her two thousand pounds more to satisfy her creditors. Whoever examines the privy purse expenses of this queen, will find no lack of economy in her own person; "her gowns were mended, turned, and new-bodied; they were freshly trimmed at an expense of four-pence, they were freshly hemmed when beat out at the bottom. She wore shoes which only cost twelve pence, with latten or tin buckles*." But her life was spent in acts of beneficence; and the rewards she bestowed on her poor, affectionate subjects, who brought her trifling offerings of early peas, cherries, chickens, flowers, &c., were very high in proportion to what she paid for her shoes. Yet, during the reign of this thrifty queen, who paid her tailor, Robert Addington, sixteen pence for mending eight gowns of

* Miss Strickland.

divers colors, at two pence apiece, the scarfs, hoods, coifs, and cauls of the ladies, were bordered with gems and pearls.

When, in 1475, King John of Aragon was endeavouring to succour the town of Perpignan, besieged by the French, he had not a jewel left—all had been pawned. Such was his penury, that he was obliged to pawn his robe lined with martin, to pay the muleteers who conveyed his effects.

Henry VIII. pawned jewels to the amount of twenty thousand crowns.

Anne Boleyn pawned her jewels just before her marriage; for, though her regal lover allowed her what were then accounted extravagant sums for her expenses, she was always in debt.

James VI. was indebted repeatedly to his jewels for resources before he inherited the English crown. But it was during the reigns of the English Stuarts that jewels were most frequently used as a resource in pecuniary straits.

In the beginning of the civil war, Henrietta Maria, the wife of Charles I., went to Holland to sell or pawn her jewels, in order to raise funds to defend him. She obtained of their High Mightinesses at Rotterdam, a sum of forty thousand guilders, and of their bank, a farther supply of twenty-five thousand guilders; of the bank at Amsterdam, three hundred and forty-five thousand guilders; of Fletcher and Fitcher, merchants at the Hague, she borrowed one hundred and sixty-six thousand guilders; on her pendant pearls she borrowed two hundred and thirteen thousand three hundred guilders; on six rubies, forty thousand guilders; and, altogether, she raised upwards of twenty thousand pounds sterling.

The exiled James and his queen had succeeded in carrying away a great many of the crown-jewels, as well as those which were their own personal property. Queen Mary Beatrice pawned and sold one by one her personal jewels, of which she had a rich store, for the support of her husband's cause, and, after his death, to relieve the wants of their exiled adherents, until she had but her marriage and coronation rings left. She has been known to take out the diamond studs from her cuffs, and send them to be sold, to relieve the imperative necessities of some loyal friend.

The jewels thus pledged were never redeemed, and probably the majority of these interesting mementos are in the hands of persons perfectly ignorant of their great historical value.

The more tranquil and prosperous reigns of the sovereigns of the House of Hanover afford no instances of English royalty reduced to such straits, at least, openly. In France the last instance known is that of the Regent, pledged by Napoleon I. to the Batavian government, for the funds required just after the bold stroke of the 18th of Brumaire.

When Murat took refuge in Corsica, after the fall of the Empire, he had but ten thousand francs left. (The treachery that had just despoiled the ex-king of Naples of two hundred thousand francs is well-known.) This sum he carried about his person in a belt. But the band around his hat was worth ninety thousand francs; one of his epaulets, fifty thousand francs; and, in addition to the gold pieces, his belt contained two diamonds valued at two hundred thousand francs;—one of these was pawned for ninety thousand francs during the owner's short sojourn at the residence of General Franseeschetti. Thus the taste for finery, for which the fugitive prince had been noted in the days of his prosperity, provided him with the means of subsistence when Fortune frowned.

Although, at the present day, the great powers of Europe have at their command other resources to meet any extraordinary demand upon their exchequers, still their crowns and personal jewels are, now and then, called upon to make up a deficiency. Such cases, however, occurring near or within our own times, cannot be made, even when known, the subjects of historical investigation, and must be left to the pens of future and untrammelled chroniclers.

Were a peep allowed behind the thick veil with which the Mont-de-Piété discreetly conceals its transactions, strange and piquant would be the mysteries revealed—mysteries, the key of which may, perchance, be also found in the hands of sundry equally discreet diplomatic agents.

But, though the Mont-de-Piété remains mute with regard, not only to the exceptional cases that would rejoice the ears of the amateurs of court scandals, and is equally careful of the secret of its humblest borrower, it publishes a yearly report of the amount of the transactions carried on there. Uninteresting as these columns of figures may at first appear, they are not without their eloquence. They indicate the extent to which luxury has arrived, and the quantity of diamonds and precious stones disseminated among the middle classes. The sum total of the loans on jewellery yearly is no less than four and a half million francs. As only two-thirds of the value at most is ever lent on an article, the value of those pledged may be safely computed at seven million francs. These loans are divided into two chief classes; those above five hundred francs each, the total of which is two million francs; and those below five hundred francs, the total amount of which is two million and a half francs.

Could the events be known that have accumulated this mass of jewels within the precincts of this vast storehouse, a curious insight would be obtained into the ways and means of Parisian life. High-born and lowly misery elbow each other on the threshold, and the dissolute artisan who enters at the common door to pledge his tools, boldly stares at the shamefaced *lion* who rings at the door of the *cabinet particulier*, where he leaves his watch.

The despair of vice and that of virtue alike find a refuge here. The secret agent of a prince in difficulties brings the time-hallowed diamonds of a regal line; the convenient emissary of the *agent de change*, or of the dissipated scion of nobility, deposits the lately-bought diamonds of the Chaussée d'Antin, or the hereditary jewels of the Faubourg St. Germain,—which a wife or mother willingly parts with to meet the note due, or pay the debt of honor, or perchance one contracted under circumstances far more wounding to the heart of the gentle owner, did she know them. Or again, the pert soubrette of some frail fair one pressed by ungallant creditors, brings the jewels offered by love and forfeited by extravagance. Here, too, may be met the widow, the daughter, reduced from affluence to penury, and compelled to sacrifice the last gift of affection, or even the cross of honor won on the field of battle by the loved and lost.

The Mont-de-Piété at Paris, established in 1777 by a royal ordinance of Louis XVI., was so successful, that it often had in its custody forty casks of gold watches that had been pledged.

In 1783, a sure indication of the depressed state of finances throughout the kingdom was apparent in the fact, that the Mont-de-Piété was literally encumbered with jewels; diamonds had so greatly diminished in price that the finest could not bring more than one-fourth of their original value.

However useful this institution may be to all classes, from the princess who sends there her coronet, to the laundress who pawns her flat-irons, it is scarcely advisable, even at the low rate of interest exacted, to leave articles too long at the Mont-de-Piété, as the following example will prove.

An actress, once celebrated for her personal charms still more than for her dramatic powers, and whose fame was at its apogee during the first empire, deposited in the

Mont-de-Piété, diamonds the gift of a sovereign, to the amount of four thousand pounds. Unfortunately the lady was as extravagant as she was beautiful; she continued her course of reckless prodigality without reflecting that time would lessen her attractions; that the public had no mercy for its superannuated favorites; that crowned admirers were growing scarce; and, what was of still more importance, that too protracted a loan would prove ruinous, as the interest would at last exceed the capital. She applied no portion of her ample receipts to lessen her debt; the consequence was that, after a period of ten years, during which the interest had amounted to the sum lent, the diamonds were sold.

CHAPTER IX.

GREAT JEWEL ROBBERIES.

Robbery of the Garde-Meubles.—Of the Diamonds of the Princess of Santa Croce.—Of Madlle. Mars.—Of the Princess of Orange.

"A diamond gone, cost me two thousand ducats in Frankfort."

"Their graces serve them but as enemies."

"'*Your virtues have proved sanctified and holy traitors to ye,*'
And that which was your proudest boast, has served but to undo ye."

Of all the precious articles that constitute the representative signs of wealth, jewels are those which contain the greatest amount within the smallest compass. Hence the ease with which they may be concealed from the most searching eyes, and conveyed mysteriously and easily from place to place. These advantages, peculiar to diamonds and precious stones, have rendered them objects of irresistible temptation to thieves in all ages and all nations.

The instinct of theft awakened by jewels is as powerful in the demi-savage slave, who secretes the brilliant he has just drawn from the bowels of the earth, as in the crowned robber, who invades his neighbour's dominions to seize a Koh-i-noor, or the child of a perverted civilization, who robs her friend of her *parure*, and ends by poisoning her own husband.

Women, who remain strangers to several categories of crimes, too often cede to the feverish longing which diamonds excite in natures inclined to evil. There have been few thefts of jewels in which a woman has not been implicated.

It is probable that such thefts would be much more frequent were it not that the difficulty of getting rid of, or using personally, purloined valuables of this nature, imposes a salutary check. Their illegal possession is attended with more danger and trouble than the robbery itself, especially in the present day. The rapidity with which electricity and steam convey information from one point to another; the press, which spreads instantly, far and wide, the description of the stolen property, paralyze the results of the crime, the fruits of which remain sterile in the hands of the perpetrator.

The piercing eye of justice has not, however, always penetrated the mystery of great jewel robberies, nor has its arm always reached the guilty. Sometimes the secret, from motives of family love and pride, remains buried in the hearts of the losers themselves. Chance, extraordinary prudence, and the honor proverbial among rogues, have, in some rare cases, rendered detection impossible.

The annals of crime record some jewel robberies, which, from the extraordinary boldness and cunning with which they have been executed, the value of the articles or the position of the owners, and sometimes from the assemblage of all these circumstances, have greatly excited public attention.

The end of the last century and the beginning of the present one have witnessed several great robberies of diamonds, not indeed presenting the romantic incidents and the celebrity of the parties concerned in the affair of the Necklace, but accompanied by circumstances that have given them the importance of historical events, and left a lasting impression on the minds of contemporaries. A brief recapitulation of a few of the principal ones may prove interesting, and, at the same time, a warning to the possessors of rich collections, and remind them of the precautions necessary to guard their treasures.

The robbery of the Garde-Meuble, so called from the

building in which the crown-jewels and regalia of France were kept, occurred in the year 1792, and is one of the most important ones known. The veil that covered that singular affair has never been raised; and, after a lapse of more than half a century, it still remains a mystery open to every conjecture, and affording ample scope to fertile imaginations.

Before the Revolution, the crown-jewels were exposed to public view, on the first Tuesday of each month during the spring and autumn. After the sanguinary days of the 1st of August and 2nd of September, 1792, it was deemed prudent to close the deposit, and the commune of Paris caused seals to be affixed to the cases containing the crown, the sceptre, the hand of justice, and other insignia; the golden chapel, enriched with diamonds and rubies, bequeathed by Cardinal Richelieu to Louis XIII.; the famous golden vessel, weighing one hundred and six marcs; and a great number of other precious articles, an inventory of which, forming a thick octavo volume, had been drawn up by virtue of a decree of the Assemblée Constituante, issued in 1791.

A man of the name of Sergent, and two other *commissaires*, had been appointed by the commune to watch these treasures. So careless or so confiding were these republican keepers, that one morning they discovered that, during the night, the colonnade surrounding the Garde-Meuble on the side towards the Place Louis XV., had been scaled, and the rooms entered through a window. The cases that had contained the crown jewels were empty; but, though the seals of the commune were broken, the locks had not been picked. Many suspected persons were arrested without any result, and the police were beginning to despair of ever obtaining a clue to the discovery of the theft or the thieves, when an anonymous letter to the commune revealed that a portion of the booty was buried in a ditch of the Allée des Veuves, in the Champs Elysées;

this alley was then one of the most lonely and worst-famed spots in Paris. Search was immediately made there, and two articles, the famous diamond, *The Regent*, and the onyx vase, called *The Chalice of Abbot Suger*, were found. The impossibility of disposing of articles so precious, and so generally known, had determined the thieves on this restitution. They sacrificed a portion to secure the remainder.

The judicial investigations were not, however, discontinued; but no better result attended them. The more mysterious the affair, the more curiosity it excited. Innumerable and most groundless were the conjectures formed. Each political party imputed the crime to its opponents. According to the royalists, Petion and Manuel had stolen these valuables in order to bribe the King of Prussia, and obtain from him the evacuation of Champagne; the republicans, on their side, accused the royalists, saying the jewels had been used to support the army of the émigrés.

The only light, and that a very doubtful one, thrown by time on this dark affair, was obtained during the trial of several individuals, charged, in 1804, with having forged bank-notes. One of the accused, who concealed his true name under the fanciful one of *Baba*, hoping to save his own head from the impending doom, made, in open court, the most complete revelations with regard to the forgery. He ended his harangue, to the great amazement of the judges and audience, with the following declaration:—"This will not have been the first time that my revelations have been of service to the public; and, if I am condemned, I will appeal with all confidence to the mercy of the emperor; had it not been for me the emperor would not be on the throne! To me alone is due the success of the battle of Marengo! I was one of the thieves of the Garde-Meubles. I had assisted my comrades to bury in the Allée des Veuves the Regent and other well-known articles, the possession of which might have betrayed them. On the promise that I should be pardoned—a promise that was faithfully kept—I revealed the hiding-place; and you are well aware, gentlemen of the court, that this magnificent diamond was pawned, by the first consul, to the Batavian government, to procure funds of which he was exceedingly in want after the eighteenth Brumaire."

Baba was again pardoned, and perhaps then made full revelations of the robbery of the crown diamonds; but, if he did, they never were made public, and were probably too late to assist justice to recover the effects or punish the guilty.

A man was still living two or three years ago, a convict in one of the French *banyues*, who was wont to boast that he had been concerned in the robbery of the greatest of the crown jewels. Strange to say, the importance of the article stolen had been a title to the respect of his lawless companions, and given him authority over them. He was looked up to as *the man who had stolen the Regent!*

Among the jewels which, on that occasion, disappeared so mysteriously and for ever, was the unrivalled Blue diamond.

Three years before the confessions of Baba had reawakened public attention to the pillage of the Garde-Meubles, another diamond robbery had caused a great sensation in the higher circles of Parisian society, as much on account of the value of the theft as of the position of the parties concerned. The diamonds were estimated at three hundred thousand francs, and were stolen from their owner, the Princess of Santa Croce, in 1801.

The princess was a Neapolitan, the widow of a Roman prince. Having joined the French party in Italy, she had been obliged to seek a refuge in Paris during the temporary reverses of the French troops in her native country.

The immense fortune of the princess, and her political

connections, had made her salons the rendezvous for her countrymen. Among the persons who composed this little court was a certain Madame Goyon des Rochettes, a Frenchwoman, but married, or supposed to be married, to the Count de Lamparelli, a Sicilian. It was as the Countess Lamparelli that she occupied the post of lady-companion to Madame de Santa Croce. One evening, while with the princess at the opera, she was unfortunately noticed and recognized by the Marquis of Loys, a recently-returned *emigré*.

The marquis coveted, at one glance, the charms of Madame Lamparelli and the diamonds of the princess. Having insinuated himself into the good graces of the young woman, when he deemed her sufficiently under his control, he boldly proposed she should rob her benefactress, in order, as he said, "to secure to both a life of love and happiness in a foreign land. The loss of these superfluous ornaments," added the marquis, to conquer the scruples of his mistress, "will be of little consequence to the princess, who, by the victories of the First Consul, will shortly be repossessed of her vast domains."

While the infamous Loys was impelling the weak countess to her ruin, he was seeking the co-operation of two professional thieves, and the crime was perpetrated one evening that the princess was dining at the Spanish embassy. Every effort of the police was for some time exerted in vain; no trace of the thieves or property could be found.

The marquis, emboldened by impunity, had the impudence to address the following anonymous note to the princess:—

"Time, signora, matters little in the affair. Yet I spent a great deal in the execution of the little trick I have played you; console yourself, however, your *patriotism* remains to you.

"The Unfindable."

The object of this note was to suggest the belief that the deed had been dictated by political revenge; nor is it improbable that this feeling was mingled with the perverse instincts of the marquis: at that time, persons who were in other respects strictly honorable and scrupulously delicate, made a merit of robbing the mails, and of carrying off by force of arms the funds of the state, on the high road; nobles did not think it derogatory to act as highwaymen.

The police was then very differently organized to what it had been in the days of anarchy and confusion, when the crown-jewels were stolen, and nothing escaped it. One of those fortuitous and apparently frivolous circumstances which frustrate the most cunning plans, betrayed the guilt of the Marquis de Loys. As a man of the world, and an amateur thief, he had fixed his attention only on the very valuable articles; but, for his accomplices, the professional gentlemen, every trifle had its value, and they had not thought it beneath them to carry away two ells of gold lace which happened to be with the diamonds. These bits of gold lace were included in the inventory of the stolen jewels, which had been printed and profusely distributed. Forgetting this, the accomplices of the Marquis of Loys, desirous that nothing should be lost, attempted to dispose of them to a fringemaker, who, finding the lace correspond with the description, had the men arrested: to shield themselves, they revealed the complicity of the marquis and countess. The countess, apprised in time, took refuge in a small provincial town, whence she might perhaps have escaped abroad; but the Count Lamparelli, either convinced of her innocence, or wishing to rid himself of his honest wife, wrote to her in urgent terms to return and vindicate her conduct. In the hope of facing the matter out, the young woman returned. She had, however, reckoned too much on the silence of her tempter; his circumstantial confession left her no chance of escape; he was condemned to twelve years at the chain-gang, and she to as many of solitary confinement. A jeweller of the Palais Royal suffered the same penalty for having received the stolen articles: in fact, he richly deserved his fate, having found means to rob the thieves themselves. When they brought him the diamonds, he had examined them carefully, and asserted that the finest stone was nothing but an imitation. To prove his assertion, he broke up before their eyes a false stone which he had artfully substituted for the real one.

The next jewel robbery of any importance, was that perpetrated at the expense of Mademoiselle Mars. The Parisian public was well acquainted with the beauty of the ornaments this celebrated actress was accustomed to display in her chief *rôles.* Attention was frequently called to them by the announcement in the papers, that "*Mademoiselle Mars would wear all her diamonds.*" This sort of exhibition was calculated to sharpen the wits, and stimulate the zeal of the professional light-fingered gentry who make theatres the arena of their feats; yet they had nothing to do with this exploit,—it was dexterously performed by amateurs.

On the 19th of October, 1827, the charming actress was to dine at a friend's, M. Armand; her host was an actor at the Théâtre Français, who played with her in almost all her pieces. Mademoiselle Mars left her hotel in a carriage, towards half-past six in the evening. At that time a number of charming little hotels had been recently built in the Rue de la Tour des Dames, for some of the artistic celebrities of the day; Talma, Mademoiselle Duchesnois, Horace Vernet, and Mademoiselle Mars, were among the fortunate owners of those elegant residences. The hotel of Mademoiselle Mars formed the angle of the Rue de la Tour des Dames and the Rue Larochefoucault. Although not a desert quarter of the town, the ground was not built upon as it is at the present day. Vacant lots were numerous in the neighbourhood; the streets little frequented, and badly lighted

Mademoiselle Mars had not left her house unguarded when she went out; a devoted female friend, an artiste who accompanied her when she acted in the provinces, remained at home that evening. The porter and his wife, who were deserving of all confidence, and her maid and valet, constituted the remainder of the household, as M. Valville, the step-father of Mademoiselle Mars, was also dining out. It was half-past nine when he returned. Constance, the maid, who knew how much her mistress liked to see her father-in-law treated with deference, was waiting for him in the porter's lodge. The eagerness with which she went forward to meet the old gentleman, to offer him her arm, and relieve him of his hat and cane, quite surprised M. Valville, who was unaccustomed to such attention from the saucy *soubrette*. The principal entrance being shut, she left him standing there, while she went round and made her way in through a back door to open the front one from within. As she admitted M. Valville, she exclaimed: "Ah! mon Dieu!—come and see, sir—they have taken everything!—the drawers are empty!—the thief must be in the house!"

Something whispered to M. Valville, that the thief was indeed in the house. He was the more inclined to think this, as Constance, in one of those unreflecting impulses so fatal to the guilty, exclaimed: "Thank Heaven! nothing can be done to me, there is no proof!" Instead of seeking the thief, who, he thought, could not escape, he deemed it advisable to inform the mistress of her loss. He hurried to Madame Armand's, and deputed her husband to break the news to Mademoiselle Mars, so as not to cause her too violent a shock.

Notwithstanding the caution he had received, Armand, albeit a good comic actor, assumed a tragic tone, and said to his fair comrade: "My dear friend, have you courage?

—you will need it!—prepare yourself!—you have been robbed of all your jewels!" "Is that all?" replied the charming woman, with most philosophical resignation, "You really made me fear a much greater misfortune." She immediately returned home, accompanied by her friends.

The officers of justice were there before her, and the depositions of the inmates were taken. It seemed impossible to fix suspicion on any one, and there was nothing to show how the thief had entered, or gone out. But the strange expression on the countenance of the maid, struck others as it had M. Valville, and her answers proving unsatisfactory, she was arrested.

Mademoiselle Mars now called to mind the eagerness manifested by Constance to enter her service; she also deeply regretted her own carelessness, in neglecting to make the usual inquiry, with regard to both the young woman and her husband, before engaging her. Those now instituted by the police brought the following particulars to light.

François Eugène Mulon, designated, even on his baptismal register, by the singular surname of *Scipion l'Africain*, was born in Paris. Having learned the art of engraving, he had gone to Geneva, and there married Constance Richard, a native of Orbe, a girl who had acquired tastes, and accustomed herself to a life of ease, little in accordance with her circumstances. To satisfy the wishes of a young wife he had formed in Paris an establishment, which, though at first prosperous, ended in their being totally ruined. Obliged to go into service, the husband obtained a situation, and the wife entered that of Mademoiselle Mars, to whom she presented a certificate which was subsequently proved to be a forgery. Mulon frequently visited his wife, and on such occasions was allowed to enter the apartment. He had no doubt seen the diamonds of Mademoiselle Mars, and well knew the place where they were kept.

Mulon was, however, supposed to be absent from Paris at the time the robbery was perpetrated. A fortnight previously he had announced rather ostentatiously, that he had just entered the service of a rich Irishman, with whom he was to travel. He had taken leave of all the servants of Mademoiselle Mars' house, and commended his wife to them during his absence. Eight days after, he was back again, saying, that his master had remained but a few hours in London, and was merely taking Paris on his way to a sea-port in the south, where he was to embark. In corroboration of this, Constance had exhibited, a few days after, a letter she said had come by post from her husband, who was in a sea-port town in the south.

Three weeks after the arrest of Constance, the Parisian authorities were notified that Mulon had been arrested in Geneva, and had confessed himself the author of the robbery. He declared that, in the frequent visits he had made to his wife, he had had opportunities of seeing the diamonds of her mistress, and the piece of furniture in which they were kept, and the sight had tempted him to appropriate them. His skill in his trade had enabled him to make the tools and false keys he needed, without the help of accomplices. Having found out that Mademoiselle Mars was to dine out on the 19th of October, he thought the moment favorable. Dressed in a blouse and muleteer's cap, and provided with a dark lantern and his implements, he had entered the apartment, gone into the bed-room, opened the secretary, and taken out the diamonds, and also two bank-notes of a thousand francs each that happened to be with them; he had, he said, left as he had entered, unperceived by any one, and had proceeded in haste to the Faubourg St. Germain. There he had taken a post-chaise, and started for Lyons, through Burgundy; at Sens, however, having met the diligence of the Messageries Royales, he had taken a seat in it, and sent the chaise back to Paris. Such was his statement.

Mulon had arrived in Geneva on the 23rd of October. He had taken all the stones from the settings, and melted these into two ingots, of the weight of forty-eight ounces. This operation he had performed at a smelter's, in whose mind it had excited no suspicion, as Mulon represented himself to be a dealer in jewels.

So far all had gone well, but he was destined to split on the rock where the majority of thieves are lost.

The gold was melted, but he now wanted to sell it. The goldsmith to whom he went for that purpose not being in, his clerk replied, that if he would leave the ingots for examination, he might return in the evening for the answer. Mulon did so, but the master having received in the interval the circular sent by the police to all the goldsmiths and jewellers of Geneva, had given notice to the magistrate of the ingots offered to him for sale by a stranger.

Mulon had no sooner entered the shop than he was questioned by the police-agent who had preceded him there. His answers were made with great presence of mind, but when he exhibited his passport as requested, the agent had no sooner glanced at it than he told him he was arrested for robbery. Though at first thunderstruck, he recovered himself and asserted his innocence, but subsequently confessed himself guilty. The diamonds he pretended he had thrown in the Rhone, but they were found, done up in a small parcel, in his boot. The stones, out of the settings, were valued at eighty-eight thousand francs, and with the ingots at ninety-six thousand francs.

On the 19th of March, 1828, the husband and wife were tried. This affair, in itself, of a private nature, was made one of public importance, such was the interest it excited throughout France. The court was thronged during the trial, the issue of which was discussed and commented upon with extraordinary eagerness. This was owing to several causes. The accomplished artiste was a great favorite with the public. The resignation with which she had heard of her loss, seemed to endear her still more. Every one was anxious to see her off the stage. This type of the perfect gentlewoman was, in the court-room, what she was everywhere, a model of grace and propriety.

But one of the great causes of excitement was the singular tendency existing, at that time, in the press and a portion of the public, to manifest more sympathy for the accused than for the accusers,—for the robbers than for the robbed,—for the assassins than for the victims. The reigning fever of that day was the abuse of certain exaggerated ideas of liberty, of respect for the defence, and of a fatalist philanthropy for criminals. The two criminals in the present case endeavoured to avail themselves of this mania; Mulon assumed the part of a rigid censor, a moralist, a husband jealous of his wife's honor. Constance, on her side, invented a fable so absurd and destitute of foundation, that nothing but the political excitement of the moment can be alleged in justification of the credence with which it was received by many.

From the time of Mulon's arrest, he persisted in asserting that he had acted entirely without the knowledge or participation of his wife; that his object in announcing that he had entered the service of a wealthy Irishman and his departure, was that he might the better watch his wife's conduct while she supposed him absent; that her exposure to all sorts of dangers in the house of an actress, rendered him painfully anxious. Unfortunately, the testimony of the witnesses contradicted him in every point. The result was, that Mulon could not possibly have entered the house on the night of the robbery; that it had been committed by his wife, and that she had passed the jewels out to him through the window of her own room;

moreover that her virtue was so perfectly safe and respected at her mistress's, that her husband, notwithstanding his cognomination of Scipio Africanus, had no reason whatever for assuming the part of an Othello.

The antecedents of Constance having been made the subject of investigation, it was found that this was not the first time she and Dame Justice had had a difference. Three years before, at the age of seventeen, while in the humble post of barmaid, in a café of the Rue St. Honoré, she had been charged by her master with stealing plate and gold and silver coins. She had, it is true, been acquitted from a want of sufficient proof; but the former accusation, and the strange and romantic defence she had made on that occasion, now militated against her. The substance of her story was probably concocted from reminiscences of the novels that had formed the staple of her literature. According to her account of herself, she was born in Switzerland, of parents who had a numerous family of children. One day the equipage of a very great lady had stopped for a moment at their door; the fine lady, charmed with the ingenuous countenance of the young girl, had requested that she might be allowed to take charge of her. To induce the parents to yield to her wishes, the stranger had placed a purse full of gold in their hands, and, while they were gazing full of surprise and joy at this little treasure, the lady had disappeared without leaving her name. Constance herself, though she remained some time in her service, had never been able to ascertain the real name of her protectress, for the noble unknown took a new one at every place she stopped at. Having made a somewhat prolonged stay at Lyons, the mysterious lady had left that city at the time of the disturbances which had been quelled by General Cannel, and gone on to Paris. On the day after her arrival, the benevolent stranger, who was only spoken of by those about her as *Madame la Comtesse*, accompanied by her young maid, drove to different parts of the town, to make purchases. They had entered a jeweller's shop, in the Rue de Richelieu, and the mistress was selecting some trinkets, when a gentleman, with a very perturbed expression, opened the shop-door, and desired to speak with the countess. The lady immediately went with the new comer to her carriage. A few words were hardly exchanged when the carriage drove off at full speed.

" Left alone in the jeweller's shop," said the accused, as she concluded her story, " I began to cry. It was in vain that I was questioned as to the name of the lady or that of the hotel at which she had been staying; I was ignorant of both, and could give no explanation. The jeweller and his wife consented to my remaining with them a short time; and they wrote to my family a letter which remained unanswered. The keeper of a café, who was a friend of the jeweller, appeared to compassionate my forlorn situation; he received me into his service, and showed me a kindness which I subsequently was made to pay dearly for, as he now persecutes me with accusations as odious as they are false."

In order to give additional interest and historical coloring to her little romance, Constance Richard had hinted that her mysterious protectress was no less a personage than Queen Hortense, and that she had thus visited Lyons and Paris to organize a vast conspiration.

Now such a fable is viewed in its true light; but at that time the public was ready to credit any absurdity, and the acquittal of the girl was pronounced amid prolonged applause, followed by a subscription, taken up on the spot, and sufficiently large to furnish the little *dot* with which she had found a husband and opened an establishment.

The new accusation brought against Constance Mulon was of too serious a nature to permit of her having recourse to such falsehoods. She took another stand, and repelled the charge and the crushing testimony with an energy and logic worthy a better cause. Her efforts to escape the punishment she had so richly deserved were vain, and the well-matched couple were each sentenced to the pillory and ten years' hard labor.

Two years had scarcely elapsed since the above trial, when every newspaper announced the robbery of the jewels of the Princess of Orange, a robbery as bold as it was mysterious, perpetrated in her own palace, and without a trace of the thieves having been found on the premises.

In this last case, neither judge nor jury was called to take cognizance of the deed. Public rumour, however, pointed to the suspected parties; though the law abstained from interfering, they were summoned before the tribunal of public opinion, and that was inexorable. But a charge thus made may be a calumny, it cannot be discussed, or rebutted, and it admits of no defence. We must, therefore, abstain from touching upon a subject which offers no elements of truth. Such of our readers as are fond of court scandals, may find ample details of this affair in several German works.

FINIS.

www.ingramcontent.com/pod-product-compliance
Lightning Source LLC
Chambersburg PA
CBHW021349210326
41599CB00011B/806
9783337009526